Studies in Smart Technologies

Advanced communication and computing technologies to develop intelligent and adaptive systems are usually referred to as smart technologies. Many of the sustainable development goals (e.g., SDG 4, SDG 6, SDG 7, SDG 8, SDG 9, SDG 11) cannot be achieved without developing smart technologies. The aim of the proposed book series, "Studies in Smart Technologies," is to provide a detailed and latest overview of the latest research findings in the field of smart technologies and allied areas. The series covers studies related to smart cities, IoT, smart sensors, AI, soft computing, and other related areas. The content published in the series will be an important reference to the policymakers, industries, researchers, practitioners, and students in computer science, electrical engineering, information technology, and aerospace engineering.

The series will publish monographs, edited volumes, textbooks and proceedings of important conferences, symposia, and meetings in the field of smart technologies and their applications.

Pravir Malik

Pioneering New Avenues in Quantum Technology

Springer

Pravir Malik
Deep Order Technologies LLC
El Cerrito, CA, USA

ISSN 3004-9881 ISSN 3004-989X (electronic)
Studies in Smart Technologies
ISBN 978-981-96-5462-8 ISBN 978-981-96-5463-5 (eBook)
https://doi.org/10.1007/978-981-96-5463-5

This Springer imprint is published by the registered company Springer Nature Singapore Pte Ltd.
The registered company address is: 152 Beach Road, #21-01/04 Gateway East, Singapore 189721, Singapore

Preface

My background is in systems thinking, and 25 years ago, I began developing a fractal theory of complex organizational systems. This work was first published as a trilogy by SAGE between 2009 and 2015, offering a primarily non-mathematical exploration of behavioral systems—spanning individuals, teams, corporations, economies, and civilizations. However, in 2015, I felt a strong need to construct a comprehensive mathematical framework to underpin this fractal perspective.

As I developed the mathematics, I discovered its applicability beyond organizational systems, extending into biological, chemical, and physical domains. This realization led to the *Cosmology of Light* series, a collection of ten books published between 2017 and 2020. The first six volumes established the mathematical foundation of fractal systems, while the final four explored its implications for quantum computation, genetics, AI, and transhumanism. Since 2021, this work has also inspired an ongoing artistic interpretation, yielding ten additional volumes so far.

The *Cosmology of Light* mathematical framework emerged in relative isolation, driven by an intuition about how the non-material manifests into material form. However, with rapid advancements in quantum computation, I felt compelled to align these theoretical insights with the evolving field of quantum technology.

Within *Cosmology of Light*, quantum computation is portrayed as a persistent, ubiquitous process in nature—where the non-material continuously materializes. This perspective suggests an inherent creativity driven by a larger systems process, in which the quantum layer is inseparable from the layers of matter and life that emerge from it. Such a viewpoint redefines the context of quantum computation and expands its perceived possibilities.

In *Pioneering New Avenues in Quantum Technology*, I take the first steps in translating this cosmological mathematics into mainstream quantum frameworks. The book outlines novel mathematics and logic, hardware architectures, algorithms, and representative applications that could serve as the foundation for a new genre of quantum technology. Many of the core ideas explored in this book have appeared in previous publications and are cited in relevant chapters. However, in this work, they have been significantly expanded and integrated into a unique quantum-based mathematical framework that far surpasses earlier versions in scope, depth, and its potential to create a foundation for pioneering quantum technologies. These ideas have also been presented in *Forbes*, as well as in *IJSSST*, *IEEE*, and *Springer Nature* conference proceedings.

The process of learning contemporary quantum mathematics, questioning its assumptions, and refining my formulations was greatly facilitated by large language models such as ChatGPT and Gemini. These tools proved invaluable in exploring intersections between existing research and my own work, and in helping to suggest easier language in some passages.

I am deeply grateful to *Springer Nature*, and especially to Aninda Bose, for their support in bringing this book to publication.

El Cerrito, USA Pravir Malik

Competing Interests The author has no competing interests to declare that are relevant to the content of this manuscript.

Introduction

The realm of quantum computing is abuzz with the potential of "probabilistic" quantum computers achieving "quantum supremacy," heralded as a significant technological breakthrough expected in the near future. This enthusiasm, driven by the phenomena of superposition and entanglement, has sparked a surge in investment and interest. However, a deeper examination suggests that the foundational premises of quantum computing require reevaluation, especially when considering the natural world's complexity that quantum computing aims to emulate.

This book explores these themes, aiming to enhance our understanding and approach to quantum technology, suggesting that the future of quantum computing lies in also leveraging fundamentally new possibilities seeded from a systems view in which the quantum level is framed as an integral part of multiple layers of matter and life, rather than merely extrapolating already envisioned ones that view the quantum level in isolation.

This book is structured into six parts.

Part I: Foundations

Part I suggests that our understanding of the basic principles of quantum mechanics in incomplete, by taking a fresh look at the double-slit experiment, exploring a quaternary fractal pattern in matter and life, and introducing a baseline mathematics for understanding quantum dynamics differently. This section challenges the conventional probabilistic view of the quantum realm, proposing an approach of qualified determinism that could better align with the observed order in nature.

Chapters include:

- Chapter 1: Surfacing a W(L) Approach to Quantum Computation
- Chapter 2: The Origins of a Fourfold Pattern
- Chapter 3: Mapping the Fourfold Pattern to Quantum Space
- Chapter 4: The Mathematics of Qualified Determinism

Part II: Reinterpretations

Part II delves into reinterpretations of some anchors foundational to quantum mechanics, offering enhanced perspectives on Heisenberg's Uncertainty Principle, Schrödinger's Wave Equation, Quanta, and Euler's Identity. This section aims to build upon the theoretical and mathematical groundwork laid in the first part, suggesting that reinterpretations of foundational anchors will allow a more comprehensive (W(L)) understanding of the quantum realm's possibilities.

Chapters include:

- Chapter 5: Uncertainty Principle and Loss of Information
- Chapter 6: Quanta and Quantum Object Code
- Chapter 7: Schrodinger's Equation and Evolution of Quantum Systems
- Chapter 8: Reimagining Euler's Endless Wave

Part III: Quaternary Interpretation of Quantum Dynamics

Part III establishes an alternative framework for understanding quantum dynamics—the Quaternary Interpretation of Quantum Dynamics (QIQD). QIQD leverages discussions through the book, tying these together in considering a simulation of a simple hypothetical atom. It also introduces the notion of 'quantum certainty' suggesting a key mechanism in changing the nature of computation.

Chapters include:

- Chapter 9: Simulation of Space, Time and Probabilities
- Chapter 10: Simulation of a Hypothetical Simple Atom
- Chapter 11: Quantum Certainty
- Chapter 12: Quaternary Interpretation of Quantum Dynamics (QIQD)

Part IV: QIQD Dynamics at Different Levels of Granularity

Part-IV leverages key aspects of QIQD to illustrate how this is responsible for the architectures of the different layers of reality from the electromagnetic spectrum through to macromolecules in cells.

Chapters include:

- Chapter 13: QIQD Dynamics at the Level of the Electromagnetic Spectrum
- Chapter 14: QIQD Dynamics at the Level of Quantum Particles
- Chapter 15: QIQD Dynamics at the Level of Atoms
- Chapter 16: QIQD Dynamics at the Level of Macromolecules in Cells

Part V: Prelude to a QIQD Design

Part V begins to lay out the philosophy of a QIQD-based quantum architecture in terms of the kinds of gates, the kind of mathematics, the kind of logic, and the QIQD algorithms required to make such a quantum computer a reality.

Chapters include:

- Chapter 17: Key Quantum Computing Principles in QIQD
- Chapter 18: Envisioning Generators and Qibits
- Chapter 19: Envisioning Other QIQD Hardware
- Chapter 20: QIQD Algorithms

Part VI: Applications

Part VI explores the practical implications of QIQD, suggesting new technologies that could arise from this enhanced understanding of quantum dynamics. This part introduces the detection of a quantum-based source of energy, the detection of emerging new structure whether at the physical, chemical, or biological levels, and a QIQD nano-based device that can that operate based on such sensing and detection. The part concludes by conceptually integrating QIQD-based quantum dynamics with the structure of spacetime as envisioned by Einstein's Field Equations, opening the door too, to QIQD-based technologies that move far beyond anything we have imagined.

Chapters include:

- Chapter 21: Detection of Organizing Quantum Energy
- Chapter 22: Sensing Emergence of New Structure
- Chapter 23: A QIQD Nano-based Device
- Chapter 24: Unifying Quantum Seeds and Spacetime

In summary, this book investigates the promise of quantum computing while reexamining its foundational principles to suggest new possibilities beyond current paradigms. By challenging probabilistic interpretations and introducing a framework of qualified determinism, it offers a novel perspective on quantum dynamics and computation. The book explores theoretical concepts, such as a quaternary fractal pattern, and proposes an alternative understanding of the quantum layer through the Quaternary Interpretation of Quantum Dynamics (QIQD). It ties these ideas to practical applications, envisioning transformative technologies for detection of quantum-based energy systems, room temperature superconductors, novel QIQD nano-devices, and others. Through a structured exploration of foundational ideas, reinterpretations, and futuristic applications, this work aims to expand the frontiers of quantum technology.

Contents

About the Author

Dr. Pravir Malik is a systems thinker, technologist, and author, specializing in quantum computing and complex adaptive systems. As the founder and technologist at QIQuantum and leader of the Forbes Technology Council's Quantum Computing community, he champions orthogonal pathways in quantum computing. Dr. Malik's fractal model links patterns across scales—from quantum dynamics to cellular structures—detailed in his ten-volume mathematical treatise, with key insights published in IEEE and Springer Nature. An author of over 25 books, Dr. Malik likens his work to that of a sculptor, using systems thinking, technology, and mathematics as artistic tools to unveil the hidden unity of knowledge and drive transformative advancements.

Part I
Foundations

Chapter 1
Surfacing a W(L) Approach to Quantum Computation

Abstract Quantum computation can be approached from distinct perspectives, illustrated by the double-slit experiment. The conventional "bottom-up" view models quantum reality from the perspective of individual photons, leading to a manufactured "r-stratum" that leverages interpreted quantum properties like entanglement and superposition for computational speed-ups and optimizations, potentially enabling qubit arrangements for simulating atoms, molecules, and even whole-chip computation. However, a "top-down" perspective views photons as constructs carrying quantum properties derived from the light source's wholeness, suggesting an alternative approach based on "l-properties" related to $W(L)$—the wholeness parsed across layers of matter and life. This $W(L)$-focused approach offers the potential for a deeper understanding of nature and the possibility of co-creation, representing a fundamentally different pathway in the pursuit of quantum computational possibilities beyond the r-stratum's engineered quantum effects.

This chapter offers an introductory exploration into generating alternative quantum computational trajectories by leveraging different interpretations of the double-slit experiment [1]. The double-slit experiment is referred to because it has played a key role in surfacing quantum dynamics at the very dawn of the 1800s, and because it has been seminal in launching inquiry into the quantum world. The commonly accepted 'bottom-up' interpretation is complemented by a 'top-down' perspective, providing insights into a functional basis for quantum computation[1] [2].

The chapter is structured to first delve into the double-slit experiment from a functional standpoint, followed by a discussion on a range of 'characters' identified through a functional analysis of the experiment. These characters serve as the foundation for developing alternative architectures for quantum computation. Subsequently, a conceptual 'double-slit world' is constructed to facilitate the exploration of the concept of 'wholeness.' This idea of building wholeness is then illustrated with practical examples from the real world. Further, the chapter examines the double-slit

[1] The idea of relooking at the double-slit from the top-down appeared in a brief note in IJSSST as noted in [2]. This note was based on a keynote delivered at a conference at Cambridge University in March 2024.

P. Malik, *Pioneering New Avenues in Quantum Technology*, Studies in Smart Technologies, https://doi.org/10.1007/978-981-96-5463-5_1

experiment from various observational perspectives to begin outlining the implications for alternative quantum computational possibilities. It also discusses potential ramifications for the design and architecture of quantum computing systems based on these insights. Finally, a parallel quantum computing developmental pathway to that which the industry has already embarked upon is suggested, which will be taken up in more detail in the remainder of the book.

1.1 Revisiting the Double-slit Experiment with a Functional Lens

Let's delve into the intricacies of the double-slit experiment. In summary, this experiment involves shining light through two slits, which results in an interference pattern on the screen positioned behind the slits. Remarkably, this pattern forms even when photons are emitted one at a time.

The experiment demonstrates a peculiar phenomenon: it appears as though each photon is aware of the interference pattern, selecting its spot on the screen to contribute to this pattern. It suggests that each photon exists in a state of superposition, where it holds multiple potential outcomes until its position is finalized by its interaction with the screen. This phenomenon also hints at quantum entanglement, where each photon is interlinked within the experiment's set up. This interpretation might be considered a bottom-up approach, aligning with the Copenhagen Interpretation of Quantum Mechanics [3]. It views each photon's behavior from an individualistic standpoint.

Conversely, a top-down interpretation sees the experiment as an indivisible entity, emphasizing a fundamental unity captured within the experiment's spacetime structure. This perspective is illustrated in Fig. 1.1, where 'W' represents a function that encompasses all attributes of light, and 'L' stands for light itself. In this model, each photon is associated with a particular attribute of $W(L)$, labeled from 'l1' to 'ln.' Each photon emitted from source $W(L)$ correlates with one of these attributes, and this relationship influences its trajectory toward the screen. In this view, photons adhere to a specific function, $b(lx)$, where 'x' ranges from 1 to 'n', and each photon's contribution to the interference pattern is dictated by this function. This interpretation does not require photons to interact across time or assume infinite possibilities, providing an alternative to the Copenhagen interpretation.

Figure 1.2 further introduces and contrasts different 'characters' that emerge from both the bottom-up and top-down interpretations of the experiment. We've already discussed $W(L)$, which represents the wholeness of light encompassing numerous quantum properties.

From the bottom-up perspective, examining the experiment involves deconstructing $W(L)$ to observe the quantum objects that manifest during the experiment. This deconstruction reveals an entity termed $r(W(L))$, a derivative function of $W(L)$, which leads to the emergence of r-properties, another key character depicted in

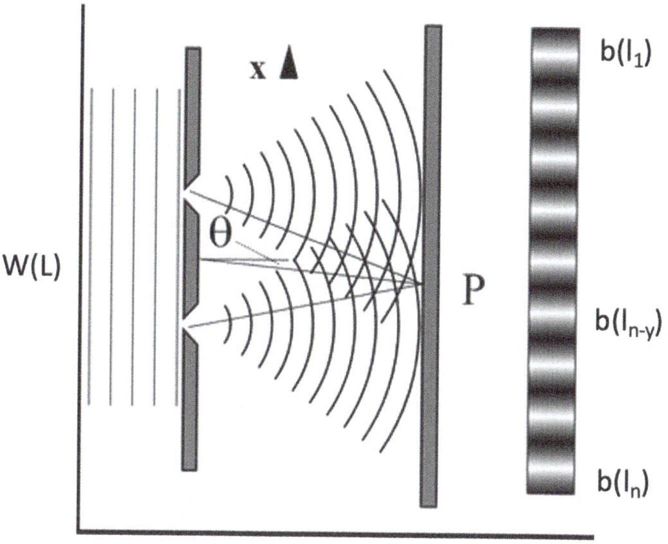

Fig. 1.1 $W(L)$ Double-slit view

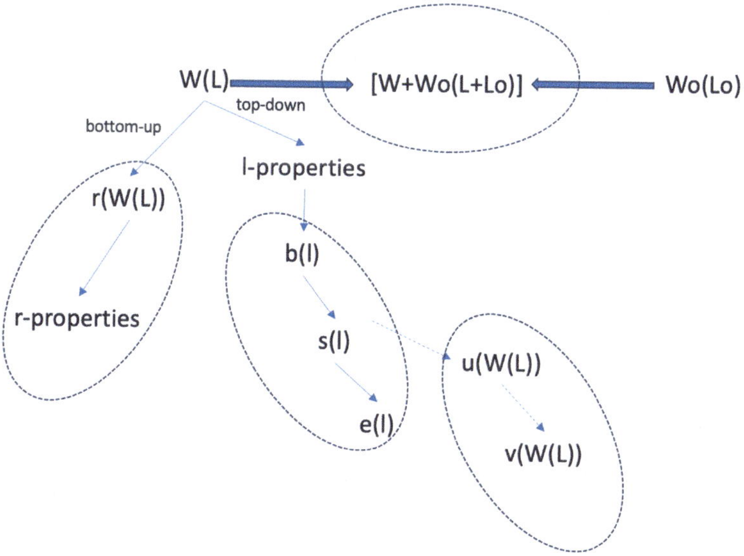

Fig. 1.2 Cast of characters

Fig. 1.2. This act of measurement, pivotal in our current approach to quantum computation, involves a process of decomposition and recomposition. We measure quantum objects, such as photons, emanating from the light source. Through measurement, we reduce the whole into measurable components and then use probability and statistics to recompose these measurements into a probability-based waveform. This waveform, while mathematically resembling the double-slit interference pattern, fundamentally differs from both the interference pattern observed at the experiment's conclusion and the potentialities within $W(L)$, as will be explained later.

A manufactured qubit, operating within the realm of measurable quantum objects, thus functions based on $r(W(L))$ and its resultant r-properties. It deals with a different set of properties than those inherent in $W(L)$, leading to a distinct computational substrate, referred to as an 'r-substrate.'

Turning to the top-down interpretation, $W(L)$ inherently possesses a series of properties, labeled l-properties. Displayed on the screen in Fig. 1.1 is another character, $b(l)$, which relates to these l-properties. Additionally, $s(l)$ represents a "seeded" layer derived from $b(l)$, and $e(l)$ is an evolutionary layer. This sequence—$b(l)$–$s(l)$–$e(l)$—suggests a different type of wholeness than that of $r(W(L))$, as it involves a progressive buildup of elements within $W(L)$. Moreover, the character $u(W(L))$ can provide a foundation for the emergence of another entity, $v(W(L))$. Separate entities, Wo(Lo), are thought to interact with $W(L)$, and a synthesis of these entities—[(W + Wo)(L + Lo)]—represents a novel conceptual unity through this combined whole.

These concepts and entities, representing various aspects of the double-slit experiment, will now be explored in further detail.

1.2 Constructing a $W(L)$-Based World

In this section, we explore a conceptual model where the entirety of a quantum system, denoted as $W(L)$, influences every aspect of its components. In this model, photons are not merely particles but carriers of the properties inherent in $W(L)$, which is defined by a unique set of quantum properties, represented as {l1, ..., ln}. Each photon emitted from $W(L)$ is intrinsically linked to one of these specific l-properties. Upon encountering the screen behind the double slits, each photon manifests as a pattern described by $b(l)$. Thus, every photon adheres to a function deriving directly from $W(L)$, and each pattern on the screen represents a distinct facet of $W(L)$.

Visually, in our thought experiment, each pattern on the screen, as depicted in Fig. 1.1, reflects a different element of $W(L)$. The presence of photons within each pattern showcases the complex functionality of each $b(l)$. Practically, each pattern could comprise countless photons, each highlighting a subtle aspect of its corresponding set.

Envision this set up where the patterns of $b(l)$ form the foundational layer in a world governed by double-slit principles. While $b(l)$ captures aspects of $W(L)$, it cannot encapsulate the whole. Yet, through combinations of these patterns, seeds are formed within a new layer—$s(l)$—each seed being a unique amalgamation of elements from the $b(l)$ patterns. This layer serves as a scaffold that begins to reveal more of what $W(L)$ contains.

Building on this, the next layer, $e(l)$, evolves from its predecessors, aiming to unfold even more complexities of $W(L)$. This evolving layer incorporates and builds upon the combinations and interactions of its foundational layers, pushing the boundaries of what can be visualized and understood about $W(L)$ in our double-slit conceptual world. Each subsequent layer complexifies the physical form that becomes a vehicle or switch for grounding more and more sophisticated function.

Now the question is how might such a world show up practically? Employing the concept of 'reverse extrapolation' can illustrate a world mirroring the holistic nature constructed in the double-slit experiment analogy. By starting with the molecular composition of living cells [4], we see a foundation built upon proteins, polysaccharides, nucleic acids, and lipids. Each molecular plan plays a distinct role: lipids organize cellular compartments enhancing functional harmony; nucleic acids, building DNA, serve as vast libraries of genetic information; polysaccharides fuel cellular activities with energy; and proteins, ubiquitous within a cell, facilitate numerous cellular functions, embodying service and presence. Collectively, these molecules exemplify a quartet of functionalities: presence, power, knowledge, and harmony.

Expanding this view to the elements of the Periodic Table [5], we identify similar functionalities among the groups: the *s*-Group elements like hydrogen and helium, essential in stellar nuclear fusion, represent power; *d*-Group metals such as iron and copper, fundamental in building infrastructure, symbolize service and presence; *f*-Group elements, with their collective atomic structures, embody harmony; and *p*-Group elements, foundational to both biological and technological structures, carry the potential for knowledge.

Further descending to the quantum particle level, we find a parallel classification [6]: bosons (facilitators of particle interactions, signifying harmony), quarks (defining elements' properties, thus carriers of knowledge), leptons (such as electrons, associated with energy release, indicative of power), and the Higgs-boson (integral to mass, representing presence).

Through 'reverse extrapolation', we trace these functional themes of presence, power, knowledge, and harmony from the quantum scale up to cells through elemental chemistry.

Illustrated in Fig. 1.3, this pattern reveals that each subsequent layer of matter and life magnifies these core functionalities. For instance, the diversity of proteins in a human cell vastly amplifies the presence function, demonstrating how the quantum-pattern of presence-power-knowledge-harmony continuously reveals more of the inherent whole without fully disclosing its entirety. This fractal [7] pattern suggests that as we observe the layered complexities of life and matter, a consistent functional pattern emerges progressively unveiling more of its quantum origins.

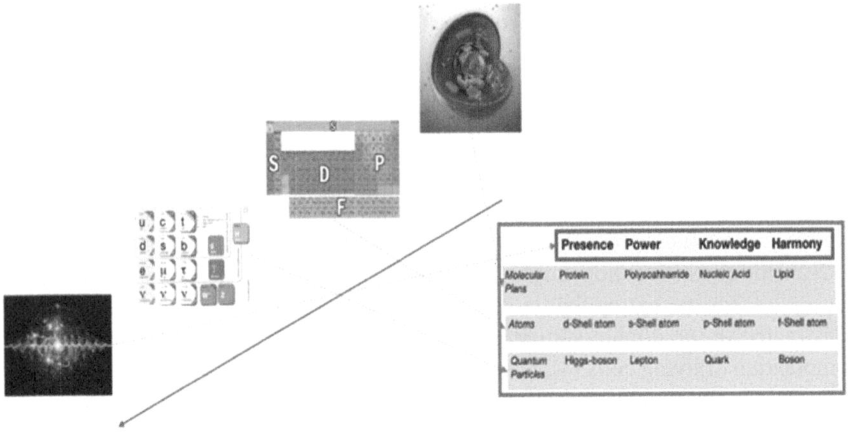

Fig. 1.3 Employing reverse extrapolation to fractal patterns

1.3 Exploring Different Observer Perspectives

Assuming the possibility of a quantum computational whole, a different architectural approach is necessary to effectively harness its capabilities. This architecture would fundamentally differ in its mathematical, logical, and structural aspects, focusing on $b(l)$, $s(l)$, or $e(l)$, where '*e*' represents an evolving functional form that continually reveals more of $W(L)$.

To further understand these nuances, insights from various observers [8] in the double-slit world are considered, which will aid in developing appropriate instrumentation and architectural design.

Observer One Scenario

This scenario mirrors the current quantum computational strategies that utilize probability and statistics to deconstruct and reconstruct $W(L)$. However, this method alters $W(L)$ into $r(W(L))$, which is a fundamentally different entity from the original whole. Despite retaining the illusion of interacting with $W(L)$, we're actually engaging with a modified entity containing a different set of properties, *r1* through *rm*. This scenario bases quantum computing capabilities on these *r*-properties, a distinct layer from the potential of *l*-properties, leading to unique superposition and entanglement phenomena.

Observer One represents the prevalent method in quantum computation that hinges on using statistical tools to break down and then attempt to rebuild $W(L)$. This process, essentially an observational act, alters $W(L)$ into $r(W(L))$, which is depicted on the left branch in Fig. 1.2. This transformed version strays significantly from the original holistic entity of $W(L)$. While $W(L)$ in its entirety still exists, our focus shifts to interacting with a different entity, $r(W(L))$, which houses a distinct set of properties, $r1$ through rm, diverging from the original l-properties ($l1$ through ln) of $W(L)$.

In the realm of practical quantum computing, manufactured qubits [9, 10]—quantum analogs of digital bits—are tailored from various quantum objects like ions, electrons, and photons, which exhibit characteristics of the r-set, not the l-set. The industry's aim is to develop a layer of quantum architecture based on these r-properties to enhance computational power significantly. One strategy, whole-chip computation, involves assembling a critical mass of qubits to model and solve complex, real-world problems through the principles of superposition [11] and entanglement [12]. Another strategy suggests using qubits to mimic quantum mechanical processes exhibited by atoms and molecules [13, 14], an approach that aligns with Dr. Richard Feynman's vision [15] of emulating natural processes from the quantum level upwards.

However, these approaches, while promising, often overlook the integrity of the quantum whole—such as maintaining the unique properties that define elements like silver or the specific functions of proteins. This oversight can be attributed to the fundamental limitations of our observational methods:

- The act of observing transforms $W(L)$ into $r(W(L))$, altering its natural state.
- Even though $W(L)$ as a whole remains intact, our attention is diverted to the realm of $r(W(L))$.

$r(W(L))$ manifests with a new set of properties $\{r1, r2, \ldots, rm\}$, distinct from the original l-properties.

- Because of the quantum mechanical nature of qubits and the processing units they are placed in, the elements of $r(W(L))$ are intertwined through superposition and entanglement, creating a new quantum state en$\{r1, r2, \ldots, rm\}$.
- Our quantum computing efforts are thus confined to exploring the capabilities within $r(W(L))$, bypassing the broader potential of $W(L)$ and excluding the possibilities of $e(l)$.

Challenges such as decoherence, scalability, and stability [16] in quantum computing may parallel the theoretical constraints in measuring and understanding the very fabric of space and time. It is postulated that an infinite amount of energy would be required to fully discern the minutiae of space and time [17], suggesting that our measurements are inherently limited by the emergent nature of these dimensions. Similarly, no matter how much energy or effort we invest in overcoming quantum computational challenges, we might never fully succeed if we continue to rely solely on the limited r-set derived from our current observational techniques, neglecting the more expansive l-set.

Observer Two Scenario

Here, observation might focus on r-properties, overlapping with Observer One, or it might uncover l-properties, tapping into the unexplored fullness of $W(L)$. Achieving focus on l-properties would require a measurement philosophy that allows for $e(l)$ to emerge, representing a significant area for research. This could lead to interacting with a more complete aspect of $W(L)$, identified as $u(W(L))$.

Observer Three Scenario

In this scenario, the focus decisively shifts to *l*-properties or the *l*-set. The observational approach in the double-slit experiment would involve measurements that allow quantum objects to demonstrate their inherent correlation with l-properties upon impacting the screen, leading to a fuller expression of $W(L)$ and potentially progressing toward $v(W(L))$.

Observer Four Scenario

This scenario considers the intersection of different wholes. Introducing another whole, Wo(Lo), perhaps influenced by a specific intent or emotion, shifts the quantum landscape. Documented experiments suggest that 'reality' might be malleable to such influences [18, 19], leading to a combined state $(W + \mathrm{Wo})(L + \mathrm{Lo})$, as depicted in the discussion. This calls for a completely different quantum computational architecture, sensitive to complex systems and adaptable to feedback and intent.

These nuanced interpretations necessitate diverse quantum computational architectures, extending beyond the conventional approach of Observer One. By exploring Observers Two, Three, and Four, new realms of quantum computational potential are unlocked, driven by a progressive understanding of $e(l)$ and aiming for a more complete realization of the quantum whole represented by $W(L)$, $u(W(L))$, and $v(W(L))$.

The four observer perspectives present diverse routes for considering alternative quantum computational architectures.

1.4 Mapping Observer Perspective to Quantum Computational Approaches

Observer One adheres to the traditional quantum computational methodology where quantum objects are analyzed and manipulated using statistical and probabilistic tools. Here, $W(L)$ is broken down and subsequently reconstructed, allowing for enhancements in computational speeds through the exploitation of r-set characteristics such as superposition and entanglement.

For Observer Two, the focus shifts toward capturing more of $W(L)$'s essence, potentially leading to $u(W(L))$ instead of settling on $r(W(L))$. This approach might necessitate postponing the measurement or altering its method significantly. The envisioned quantum computer would fundamentally differ, relying on a $b(l)$-stratum. Such a structure would support manipulation based on 'function' rather than on the properties of quantum objects within the *r*-set. A possible design with illustrative mathematics that will drive such a computer is illustrated in Fig. 1.4. Note that the gist of the architecture and gates and the underlying math driving these gates will be further elaborated in the book.

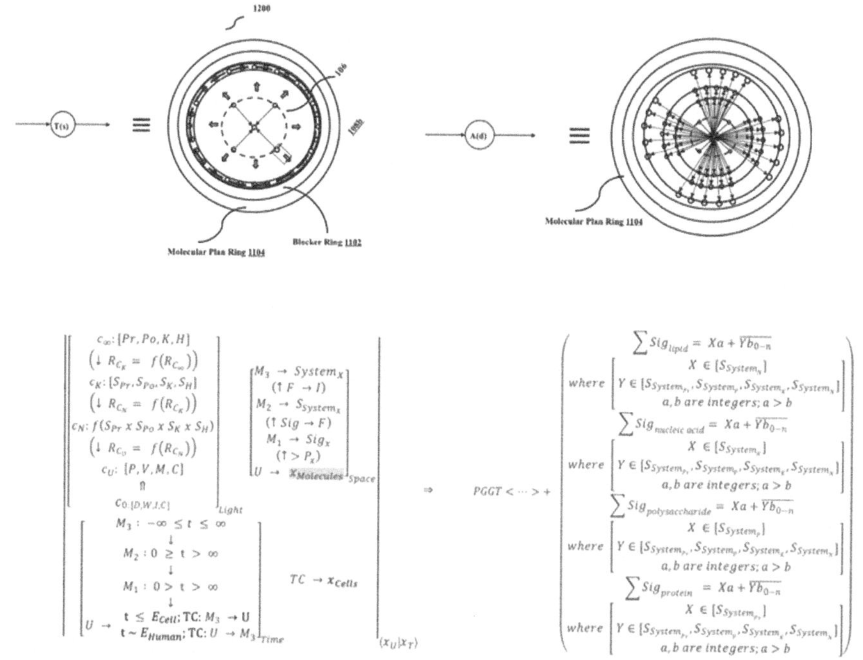

Fig. 1.4 Depiction of function—propagation gates and some underlying math

Observer Three seeks a deeper appreciation of the quantum domain's totality, drawing metaphorical inspiration from the interference patterns observed in experiments. This angle promotes a computation model that is function-oriented, altering how quantum phenomena are harnessed. Such a model would incorporate function-propagation gates in innovative layouts to push the boundaries of computational power toward $v(W(L))$. This approach aligns with and extends the groundbreaking theories of historical figures like Feynman and Schrödinger. Chapters 2–12 begin to lay out the mathematics, and Chaps. 17–20 suggests the hardware that would drive such a computational approach.

Observer Four delves into the dynamics between distinct wholes, which requires a highly sophisticated computational model capable of embracing this complexity. This model might include elements of human interaction, reflecting theories from studies on nano-cyborg technologies and views of the universe as a complex adaptive system, and are touched up in Chaps. 21–24. Such an approach could pave the way for a new paradigm in quantum computing that includes human factors as integral components of the computational process.

While we are developmentally well underway along the trajectory suggested by the Observer One scenario, the Observer Two, Three, and Four scenarios unveil additional dimensions to quantum computation, suggesting a parallel developmental roadmap to that already being pursued by the Observer One scenario.

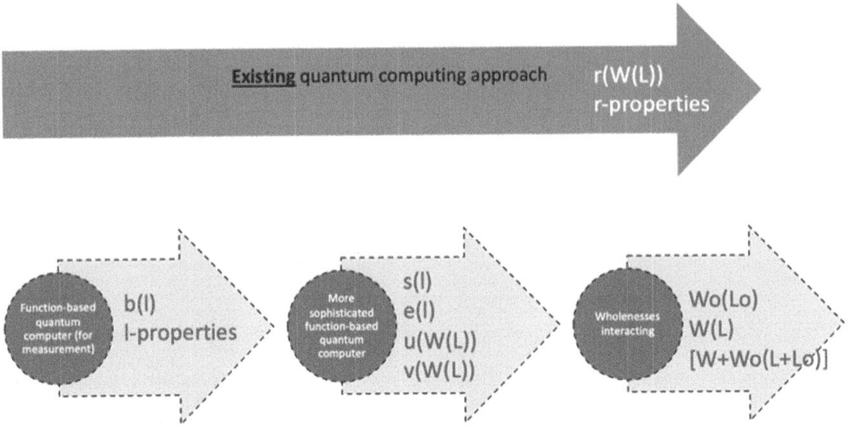

Fig. 1.5 Parallel pathways

Our current focus, in the lens set up by this chapter, is on creation of an *r*-stratum that deals with *r*-properties and is bound by $r(W(L))$. Tens of billions of dollars [20] have already been injected into this, and it is anticipated that we shall see tangible results approximately within ten years. Figure 1.5 summarizes the current approach in the upper arrow. A second approach is depicted by the series of arrows below the upper one.

The lower set of arrows start with a function-based quantum computer that aims to measure differently, so that the beginnings of an *l*-stratum dealing in *l*-properties can be constructed. This will yield a more sophisticated *l*-stratum-based quantum computer allowing the surfacing of $s(l)$ and $e(l)$ and bound by the possibilities of $v(W(L))$. Mastery over this type of quantum computer would then allow the quantum computing stratum to deal more with the possibilities bound by $(W + \mathrm{Wo})(L + \mathrm{Lo})$.

These innovative approaches and their implications will be explored in greater detail throughout the remaining sections of the book, providing an understanding of the potential shifts in quantum computational paradigms.

1.5 Further Reading

- "The Feynman Lectures on Physics, Vol. III" by Feynman, R. P., Leighton, R. B., & Sands: Feynman's lectures including that of the double-slit experiment and its implications for quantum mechanics.
- "Quantum Mechanics: The Theoretical Minimum" by Leonard Susskind and Art Friedman: This book offers a technical approach to quantum mechanics.
- "Principles of Quantum Mechanics" by R. Shankar: A textbook that provides a comprehensive introduction to quantum mechanics, including a thorough discussion of quantum dynamics.

- "The Cell: A Molecular Approach" by Geoffrey M. Cooper and Robert E. Hausman: This textbook offers a more concise overview of molecular cell biology.
- "The Disappearing Spoon" by Sam Kean: This book tells the stories behind the elements and their discoveries.
- "The Elements: A Visual Exploration of Every Known Atom in the Universe" by Theodore Gray: A visual book with high-quality photographs of each element and descriptions of their properties and uses.
- "Introduction to Elementary Particles" by Griffiths, D.: A textbook that provides a comprehensive introduction to particle physics, including the Standard Model.
- "Basics of Fractal Geometry" by Michael Frame and Benoit Mandelbrot: This research paper provides a more technical overview of fractal geometry, including its mathematical formalism and applications in science and engineering.
- "Can quantum–mechanical description of physical reality be considered complete?" by Bohr: This paper by Niels Bohr discusses the concept of complementarity and the role of the observer in quantum measurements.
- "Physics and Philosophy: The Revolution in Modern Science" by Heisenberg, W.: In this book, Heisenberg explores the philosophical implications of quantum mechanics, including the observer's role and the concept of indeterminacy.
- "Mathematical Foundations of Quantum Mechanics" by Von Neumann, J.: This classic text provides a rigorous mathematical framework for quantum mechanics and discusses the measurement problem, including the role of the observer in wave function collapse.
- "Quantum Computation and Quantum Information" by Michael A. Nielsen and Isaac L. Chuang: This is a comprehensive textbook often considered the "bible" of quantum computing. It covers the theoretical foundations, algorithms, error correction, and physical implementation of quantum computers.

References

1. Ananthaswamy A (2023) Particle, wave, both or neither? The experiment that challenges all we know about reality. Nature 618(7965):454–456
2. Malik P. Leveraging the double-slit experiment to explore new horizons in quantum computation. IJSSST. https://doi.org/10.5013/IJSSST.a.25.01.12
3. Faye J (2024) Copenhagen interpretation of quantum mechanics. Stanford encyclopedia of philosophy. Stanford University. https://plato.stanford.edu/entries/qm-copenhagen/
4. Goodsell DS. The machinery of life. Center for computational structural biology, the Scripps research institute. Accessed 13 Jan 2025. https://ccsb.scripps.edu/goodsell/machinery-of-life/
5. Royal Society of Chemistry. Periodic table. Accessed 13 Jan 2025. https://www.rsc.org/periodic-table/
6. CERN. The standard model. Accessed 13 Jan 2025. https://home.cern/science/physics/standard-model
7. Fractal Foundation. What are fractals? Accessed 13 Jan 2025. https://fractalfoundation.org/resources/what-are-fractals/
8. Bohr N (1935) Can quantum-mechanical description of physical reality be considered complete? Phys Rev 48(8):696–702. https://doi.org/10.1103/PhysRev.48.696

9. Quantum Inspire. What is a qubit?. Accessed 13 Jan 2025. https://www.quantum-inspire.com/kbase/what-is-a-qubit/
10. University of Waterloo Institute for Quantum Computing. What is a qubit?. Accessed 13 Jan 2025. https://uwaterloo.ca/institute-for-quantum-computing/quantum-101/quantum-information-science-and-technology/what-qubit
11. California Institute of Technology. Quantum superposition. Science exchange. Accessed 13 Jan 2025. https://scienceexchange.caltech.edu/topics/quantum-science-explained/quantum-superposition
12. California Institute of Technology. What is entanglement and why is it important?. Science exchange. Accessed 13 Jan 2025. https://scienceexchange.caltech.edu/topics/quantum-science-explained/entanglement
13. Bravyi S, Kitaev A (2019) Fermionic quantum computation. Phys Rev Res 1(3):033033
14. Xie W (2023) Use VQE to calculate the ground energy of hydrogen molecules on IBM Quantum. arXiv preprint arXiv:2305.06538
15. Feynman RP (1982) Simulating physics with computers. Int J Theor Phys 21(6):467–488
16. Dyakonov M (2018) The case against quantum computing. IEEE spectrum. https://spectrum.ieee.org/quantum-computing-skeptics
17. Philosophical Society of Washington. The doom of spacetime. PSW science. Accessed 13 Jan 2025. https://pswscience.org/meeting/the-doom-of-spacetime/
18. Emerging Technology from the arXiv (2019) A quantum experiment suggests there's no such thing as objective reality. MIT Technol Rev. https://www.technologyreview.com/2019/03/12/136684/a-quantum-experiment-suggests-theres-no-such-thing-as-objective-reality/
19. Emerging Technology from the arXiv (2010) Quantum entanglement can be a measure of free will. MIT Technol Rev. https://www.technologyreview.com/2010/08/25/200871/quantum-entanglement-can-be-a-measure-of-free-will/
20. Malik P (2022) Managing the quantum bubble. Forbes. https://www.forbes.com/sites/forbestechcouncil/2022/11/17/managing-the-quantum-bubble/

Chapter 2
The Origins of a Fourfold Pattern

Abstract A thought experiment involving multiple Hilbert spaces generates a root pattern reflecting wholeness, $W(L)$, and the four functional aspects of presence, power, knowledge, and harmony (P, Po, K, H). Beginning with an originating Hilbert space where light is imagined traveling infinitely fast, these four properties emerge. Quantization of this light into subsequent Hilbert spaces—first at a slower speed k, then at speed g, and finally at c (the known speed of light)—reveals increasingly complex structures. At speed c, a fourfold quantum object profile or code emerges, defined by space, time, energy, and gravity dimensions, which are shown to be symmetrical expressions of P, Po, K, and H. This fourfold code, reflected in the Standard Model, the Periodic Table, and cellular molecular plans, suggests a fundamental architectural principle for arising matter and life. This framework, summarized by four generated Hilbert spaces, provides a basis for understanding the evolution of quantum systems via Schrödinger's equation, Hamiltonian operators, and eigenvalues, and a universal wavefunction to be explored in later chapters.

In Chap. 1, we were introduced to a constructed $W(L)$-based world. It is suggested that this world shows up at the levels of matter and life, as the fourfold Standard Model [1] at the level of quantum particles, the four categories of the Periodic Table [2] at the level of atoms, and as the four molecular plans [3] at the level of living cells.

We posited that there is a root pattern at the quantum levels that shows up as ever-more sophisticated fractal [4] patterns at the subsequent levels of physics, chemistry, and biology, without ever exhausting the possibilities of the information originally present at the quantum levels.

The question then, is how might that reservoir of infinite information that informs the quantum levels be modeled? The generated model can be thought of as the conceptual or mathematical space [5] from which the fractal patterns at the levels of physics, chemistry, and biology show up.

This model will be generated by imagining light to exist at different speeds. Note that Einstein's Theory of Relativity [6] does not disallow particles from traveling

P. Malik, *Pioneering New Avenues in Quantum Technology*, Studies in Smart Technologies, https://doi.org/10.1007/978-981-96-5463-5_2

faster than light if they are already doing so [7, 8]. The issue is in accelerating from below the speed of light to the speed of light.

2.1 Generating the Root Model

Imagine that there is a light source in the center of a vast volume. This light is traveling infinitely fast, and since it is traveling infinitely fast, it is going to be present everywhere in this volume simultaneously. So, it has a property of all-presence. If there's anything not of the nature of light that arises in this volume, it's going to be overpowered by the nature of light, so this light has a property of all-power. Since everything that arises in this space is happening in the fabric of light, light has a "knowledge" of everything that's happening: this light has a property of all-knowledge. Since everything is implicitly happening in the nature of light, there's also a property of all-harmony.

So intuitively we can get a sense that associated with light are four properties that will be referred to as 'presence', 'power', 'knowledge', and 'harmony' [9].[1] This can be summarized as a Hilbert space [10], $\mathbb{H}_{L_\infty}$, of light (L) existing infinitely (∞) fast. This space can be represented by a set ($\{\ldots\}$) where 'Pr' is 'presence', 'Po' is 'power', 'K' is 'knowledge', and 'H' is 'harmony' (Eq. (2.1), *Hilbert Space of Light Traveling Infinitely Fast*):

$$\mathbb{H}_{L_\infty} = \{Pr, \mathrm{Po}, K, H\}. \tag{2.1}$$

Note that a Hilbert space offers a generalized and abstract representation of quantum systems, accommodating an infinite number of dimensions. While light as conceptualized—traveling infinitely fast—can be thought of as a Hilbert space with infinite degrees of freedom, $\mathbb{H}_{L_\infty}$, the wavefunction of interest that we will arrive at Chaps. 3 and 4 can be thought of as representing the quantum state of properties to do with viewing light in its aspect of speed. This is extremely relevant to the focus of this book since known matter and life arise where light exists at a speed of 186,000 miles per second.

As will be seen, insight into practical quantum possibilities, into the layer of quantum dynamics, and into effective architectures and logic of quantum computation, will also better arise when viewed from the point of view of $\mathbb{H}_{L_\infty}$. Such a treatment will also allow a deeper understanding of possible operations and operators, including the range of "energy-based operations" that can be subject to a Hamiltonian operator, where a Hamiltonian [11] can be thought of as an operator that generates the time evolution of a quantum system. As a result of more completely

[1] The root pattern was originally generated across the series of Cosmology of Light books referred to in [9] and then subsequently in several IEEE papers—https://ieeexplore.ieee.org/author/37086022058.

understanding the range of energies, or energy spectrum, a better understanding of eigenvalues [12] will also arise.

Such an expanded view will have a practical bearing of computations possible at the quantum level, which as this book will illustrate, are vastly different than what is currently being pursued by the quantum computing industry. The focus of this book after all is the quantum computational architecture and possibilities that arise in the '*l*-stratum' where *l*-properties based on wholeness are the natural dynamics, rather than the *r*-stratum, where *r*-properties based on reductionism are the natural dynamics. As discussed in Chap. 1, the latter stratum is the focus of the quantum computational industry.

Light traveling infinitely fast is represented by (2.1). Suppose it projects itself at slower speed though, what will happen, and how will the resulting Hilbert space be depicted? First, since the properties of light—Pr, Po, K, H—are true in $\mathbb{H}_{L_\infty}$, some symmetrical aspect of them should also be true in the resulting projected realm. In fact, any slowing down of light should make those fundamental properties more "visible". That is, it should result in a 'quantization' so that the essence of what is meant by Pr, Po, K, H becomes more clear. But intuitively we already know this because when light is traveling at 186,000 miles per second, then these four properties show up as the very structure of quantum particles, atoms, cells, and so on, as previously suggested.

To understand 'quantum' in a more conceptual way, think of it as a mechanism that collects and materializes information in those spaces generated by imagining light traveling infinitely fast. Imagine the original space where light travels infinitely fast as containing an infinite amount of information. If light then projects itself to a realm where it travels more slowly, quanta can be thought of as a bridge-mechanism that parses some of the information from the original field and translates it into a symmetrical and more material form in a target field.

At a projected speed of 'k', slower than ∞, but much faster than 186,000 miles per second (c), there is therefore a differentiation or quantization that occurs so that the properties summarized by (2.1) transform into corresponding sets of presence (S_{Pr}), power (S_{Po}), knowledge (S_K), and harmony (S_H). These sets can be thought of as containing up to infinite elements related to the essential property of each set. Hence Eq. (2.2), *Hilbert Space of Light Traveling at k*, is depicted as:

$$\mathbb{H}_{L_k} = \{S_{\text{Pr}}, S_{\text{Po}}, S_K, S_H\}. \tag{2.2}$$

With another act of quantization, mechanically due to imagining light slowing down to a speed 'g', which is faster than c but slower than k, the resulting differentiation is defined to be such that an infinite number of seeds (n) are created that would each be unique combinations of elements from each of the four property sets of Light ($f(S_{\text{Pr}}xS_{\text{Po}}xS_KxS_H)\}$). The resulting Hilbert space, $\mathbb{H}_{L_g}$, would be depicted by Eq. (2.3), *Hilbert Space of Light Traveling at g*:

$$\mathbb{H}_{L_g} = \{\ n|f(S_{\text{Pr}}xS_{\text{Po}}xS_KxS_H)\}. \tag{2.3}$$

Now if light were to be projected at the known speed of c, where materialization is known to occur, then the infinite number of 'subtle' seeds depicted by $\mathbb{H}_{L_g}$ in (2.3), would result in the macro-parameters of space, time, energy, and gravity, as per the following logic [13]: First, the subtle-seeds would become the substance of 'space'. Then the uniqueness of that which is contained in each seed will show itself materially through a process of 'time' that will be relative to each seed. In this process of relative time, a transformation will take place from the subtle to the material through the application of 'energy'. Finally, each seed will have a relationship with other seeds through the mechanism of 'gravity'. Hence, each seed will have a precise space–time-energy-gravity profile.

However, when we look at the essence of space, we know that since it has an archetype of everything that will arise, it has to do with 'knowledge'. Further, regardless of any obstacle, the essence of what's in a seed will manifest through relative time, and in its essence such manifestation can be seen as an act of 'power'. Energy transforms something that's subtle into a material form, and this has to do with 'presence'. Finally, gravity is showing an arrangement between different seeds, which in essence has to do with 'harmony'. The point being that space, time, energy, gravity is also a symmetrical face of these four properties of light.

The resulting Hilbert space, $\mathbb{H}_{L_c}$, would be depicted by Eq. (2.4), *Hilbert Space of Light Traveling at c,* where '*S*' is 'space', '*T*' is 'time', '*E*' is 'energy', and '*G*' is 'gravity':

$$\mathbb{H}_{L_c} = \{S, T, E, G\}. \tag{2.4}$$

Summarizing, in this set of Hilbert spaces, the originating Hilbert space, $\mathbb{H}_{L_\infty}$, is created where light is traveling infinitely fast, and in that infinitely fast layer there are properties that show up: presence, power, knowledge, and harmony. When that light is projected at a slower speed, or quantized, it shows up as a second subsidiary Hilbert space, $\mathbb{H}_{L_k}$, depicted by four infinitely large sets that reveals something that could not be seen in the original Hilbert space. Further quantization, due to light being projected at speed *g*, results in the Hilbert space $\mathbb{H}_{L_g}$, depicted by infinite number of unique seeds that have contributions from all the four sets in unique combinations. A final quantization due to light being projected at *c*, results in the Hilbert space, $\mathbb{H}_{L_c}$, depicted by a fourfold quantum object profile or code specified by space, time, energy, and gravity dimensions.

These dimensions are nothing other than a symmetrical face of *P*, Po, *K*, and *H*, and as will be reinforced in further chapters, architect the structure of arising matter and life, as already hinted at by the Standard Model, the Periodic Table, and the structure of living cells as made evident by fourfold molecular plans. In other words, anything that shows up at 186,000 miles per second can be said to have a fourfold space–time–energy–gravity quantum object code. This becomes important in understanding evolution of quantum systems as depicted by Schrödinger's equation [14], Hamiltonian operators that can initiate time evolution of quantum systems, and eigenvalues that quantum systems may surface. These will be taken up in more detail in subsequent chapters. The four generated Hilbert spaces, summarized by (2.1–4),

Table 2.1 Mathematically generated conceptual spaces by considering light at different speeds

#	Layer	Presence	Power	Knowledge	Harmony
–ni	Light source	{..., Pr−1, ... Pr−n, ...}	{..., Po−1, ... Po−n, ...}	{..., K−1, ... K−n, ...}	{..., H−1, ... H−n, ...}
...					
–i	Light seed	{Pr, po, k, h}	{pr, Po, *k*, *h*}	{pr, po, *K*, *h*}	{pr, po, *k*, *H*}
0	Quantum object code	Energy	Time	Space	Gravity

will also be instrumental in understanding a universal wavefunction at the heart of quantum systems, to be introduced in a subsequent chapter.

The mathematically generated conceptual spaces by considering light to exist at different constant speeds are summarized in Table 2.1.

What is important to note is that these conceptual spaces are depicted as imaginary [15] (i) layers as indicated by the '#' column in Table 2.1. Quantum object code can also then be seen as a mechanism involving reference to an 'imaginary' layer, and reference back from an imaginary layer to the '0' layer of quantum object code. Such a double action can be depicted by i^2. Mathematically, '$i^2 = -1$', and this gives us a clue that Euler's identity $\left(e^{i\pi} + 1 = 0\right)$ [16] will also need to be considered in the materializations that can be due to quantum computation and will be taken up in Chap. 8.

2.2 Further Reading

- "Introduction to Elementary Particles" by Griffiths, D.: A classic textbook that provides a comprehensive introduction to particle physics, including the Standard Model.
- "An Introduction to the Standard Model of Particle Physics" by Cottingham, W. N., & Greenwood, D. A.: Another textbook that covers the Standard Model in detail, with a focus on the theoretical foundations.
- "The Periodic Table: Its Story and Its Significance" by Eric R. Scerri: This work delves into the history, philosophy, and scientific details of the periodic table, exploring its development and impact on chemistry.
- "The Periodic Table: A Field Guide to the Elements" by Paul Parsons and Gail Dixon: This book offers a comprehensive guide to the elements, with detailed information on their properties, uses, and occurrence in nature.
- "Molecular Biology of the Cell" by Alberts et al.: This textbook provides an in-depth exploration of cellular processes at the molecular level, including gene expression, protein synthesis, signaling pathways, and more.

- "The Geometry of Meaning: Semantics Based on Conceptual Spaces" by Gärdenfors, P.: This book delves into the relationship between conceptual spaces and language with implications for philosophy of mind and cognitive science.
- "Applications of Conceptual Spaces: The Case for Geometric Knowledge Representation" by Zenker, F., & Gärdenfors, P.: This edited volume showcases a variety of applications of conceptual spaces in different domains, such as linguistics, psychology, artificial intelligence, and cognitive robotics.
- "Relativity: The Special and the General Theory" by Einstein, A.: A classic, accessible explanation by Einstein himself.
- "Introductory Functional Analysis with Applications" by Kreyszig, E.: A textbook on functional analysis that covers Hilbert spaces in detail, along with other important topics in the field.
- "Functional Analysis" by Rudin, W.: A textbook that provides a rigorous treatment of functional analysis, including Hilbert spaces and their applications.
- "Cosmos" by Carl Sagan: A classic popular science book that explores the universe and our place in it, covering topics from cosmology to evolution.
- "The Book of the Cosmos: Imagining the Universe from Heraclitus to Hawking" edited by Dennis R. Danielson: A collection of primary sources from key figures in the history of cosmology, offering insights into the development of ideas about the universe.
- "Schrödinger: Life and Thought" by Walter Moore: This biography offers a comprehensive look at Schrödinger's life, his scientific work, and his philosophical views.
- "'What Is Life?' with Mind and Matter and Autobiographical Sketches" by Erwin Schrödinger: This collection includes Schrödinger's famous essay "What Is Life?" where he explores the relationship between physics and biology.
- "Erwin Schrödinger and the Quantum Revolution" by John Gribbin: This biography focuses on Schrödinger's scientific contributions, particularly his development of the wave equation and his role in the quantum revolution. It also discusses his debates with other physicists and his philosophical interpretations of quantum mechanics.
- "The Euler Archive: http://eulerarchive.maa.org/": This is an online resource for exploring Euler's work. It contains digitized versions of his original publications, along with summaries, translations, and commentary.
- "Euler: The Master of Us All" by William Dunham: This book provides an accessible introduction to Euler's contributions to various fields of mathematics, including number theory, calculus, and graph theory. It's a great starting point for anyone interested in Euler's work.
- "The Four Archetypes: Mother, Father, Rebirth, Spirit" by Carl Jung: While not directly related to physics, Jung's work on archetypes explores the psychological significance of fourfold structures.
- "The Republic" by Plato: Plato's theory of the four elements and his exploration of ideal forms touch upon the concept of fundamental principles.

References

1. Cooper K (2023) What is the standard model of physics?. Space.com. https://www.space.com/standard-model-physics
2. Levi P (2000) The periodic table. Translated by Raymond Rosenthal. Penguin Books, London
3. Goodsell DS. The machinery of life. Center for computational structural biology. The Scripps Research Institute. Accessed 13 Jan 2025. https://ccsb.scripps.edu/goodsell/machinery-of-life/
4. Falconer KJ (2013) Fractals: a very short introduction. Oxford University Press, Oxford
5. Gärdenfors P (2000) Conceptual spaces: the geometry of thought. MIT Press, Cambridge, MA
6. Einstein A (2020) On the electrodynamics of moving bodies. Translated by John Walker. Internet Archive. https://archive.org/details/einstein-1905-relativity
7. Tachyon. Encyclopædia Britannica. Accessed 13 Jan 2025. https://www.britannica.com/science/tachyon
8. Dixon L (2016) If tachyons exist, how might they be detected?. Astronomy.com. https://www.astronomy.com/science/if-tachyons-exist-how-might-they-be-detected/
9. Malik P. 20-book cosmology of light series. Amazon Author Page. Accessed 14 Jan 2025. https://www.amazon.com/author/pravirmalik
10. Brezhnev YV (2021) Why and whence the Hilbert space in quantum theory?. arXiv preprint arXiv:2110.05932. Revised 1 March 2023
11. Cline D (2025) 18.3: hamiltonian in quantum theory. In: Variational principles in classical mechanics. LibreTexts. https://phys.libretexts.org/Bookshelves/Classical_Mechanics/Variational_Principles_in_Classical_Mechanics_(Cline)/18%3A_The_Transition_to_Quantum_Physics/18.03%3A_Hamiltonian_in_Quantum_Theory
12. 6. Eigenvalues and eigenvectors. In: Math for quantum. TU Delft. Accessed 13 Jan 2025. https://mathforquantum.quantumtinkerer.tudelft.nl/6_eigenvectors_QM/
13. Malik P (2022) The role of a light-based quantum computational model in the creation of an oscillating universe. In: 2022 IEEE 12th annual computing and communication workshop and conference (CCWC), Las Vegas, NV, USA, pp 0953–0959. https://doi.org/10.1109/CCWC54503.2022.9720904
14. Schrödinger equation. In: A dictionary of physics. Oxford University Press. Accessed 13 Jan 2025. https://www.oxfordreference.com/display/10.1093/oi/authority.20110803100446949
15. Imaginary Number. In: UXL encyclopedia of science. Encyclopedia.com. Accessed 13 Jan 2025. https://www.encyclopedia.com/science-and-technology/mathematics/mathematics/imaginary-number
16. Nahin PJ (2017) Dr. Euler's Fabulous formula: cures many mathematical ills. Princeton University Press, Princeton

Chapter 3
Mapping the Fourfold Pattern to Quantum Space

Abstract This chapter introduces a hierarchical framework of Hilbert spaces representing progressively symmetry-broken layers of quantum reality. Beginning with a source space (H_{L_∞}) of maximal symmetry, the hierarchy culminates in physical reality (H_{L_c}) where quantum objects with defined space–time–energy–gravity properties emerge. “Seeds,” introduced at an intermediate layer (H_{L_g}), act as functional mappings, encoding deterministic structures from abstract quantum properties into observable phenomena. The chapter explores the role of eigenstates and eigenvalues, their symmetry-preserving transformations under high-dimensional groups (SU(n), SO(n), U(n)), and introduces a universal Hilbert space ($H_{\text{universal}}$) encompassing all layers. Utilizing the density matrix formalism to represent the coexistence of deterministic seed properties and probabilistic features, the framework employs projection mechanisms to reduce higher-dimensional wavefunctions to observable states. Finally, the chapter briefly introduces a Hamiltonian governing seed dynamics and feedback mechanisms between layers, offering new insights into the interplay of determinism and probability in quantum mechanics and providing a foundation for exploring complex systems, emergent behaviors, and quantum computation.

This chapter introduces a novel framework that maps the evolution of quantum systems across a hierarchy of four interconnected Hilbert spaces. At the core of this framework is the universal Hilbert space, which integrates all four layers through a tensor product, enabling a unified representation of quantum systems. The chapter explores how seeds serve as deterministic blueprints for quantum objects while retaining probabilistic features in their interactions. By incorporating eigenstates, eigenvalues, and the density matrix formalism, the chapter provides a mechanism to balance deterministic structure and quantum uncertainty. This framework also introduces a universal wavefunction, operators, and projection mechanisms to describe the evolution of quantum systems and their observable properties.

P. Malik, *Pioneering New Avenues in Quantum Technology*, Studies in Smart Technologies, https://doi.org/10.1007/978-981-96-5463-5_3

3.1 Universal Hilbert Space

The previous chapter proposed four Hilbert spaces. A source Hilbert space, H_{L_∞}, is associated with a conceptual space generated by imagining light traveling infinitely fast and contains the most fundamental properties (Pr, Po, K, H). This space exhibits maximal symmetry. A derivative Hilbert space, H_{L_k}, is generated when light is imagined slowing to a constant speed 'k' and represents a symmetry-broken layer where new properties begin to emerge structured as infinite sets derived from Pr, Po, K, H. A second symmetry-broken Hilbert space, H_{L_g}, corresponds to light imagined slowing down further, to 'g'. This layer introduces "seeds," which are unique combinations of the earlier properties. A final Hilbert space, H_{L_c}, is associated with light traveling at the speed of materialization 'c'. This space reflects macroscopic phenomena, such as space, time, energy, and gravity. Note that the seeds impose internal constraints, ensuring that quantum objects have determinate properties. This introduces an important element of determinism in quantum mechanics that will be taken up in more detail subsequently.

A composite Hilbert space, $H_{\text{universal}}$, represents the construction of a universal Hilbert space by combining the four hierarchical Hilbert spaces $H_{L_\infty}, H_{L_k}, H_{L_g}, H_{L_c}$ using the tensor product ($\otimes$), as depicted in Eq. (3.1), *Universal Hilbert Space*:

$$H_{\text{universal}} = H_{L_\infty} \otimes H_{L_k} \otimes H_{L_g} \otimes H_{L_c}. \tag{3.1}$$

This is the total Hilbert space that captures the entire quantum system. It integrates the structure and properties of all four layers $H_{L_\infty}, H_{L_k}, H_{L_g}, H_{L_c}$. The tensor product ($\otimes$) is a mathematical operation used to combine Hilbert spaces [1]. If $|u\rangle \in H_1$ and $|v\rangle \in H_2$ then the combined state is $|u\rangle \otimes |v\rangle \in H_1 \otimes H_2$. The tensor product allows all four layers to coexist in a single mathematical framework. This is important for describing interactions between high-dimensional symmetries (H_{L_∞}) and emergent macroscopic phenomena (H_{L_c}) [2] and preserving information from the source space while incorporating symmetry-breaking effects.

3.2 Universal Wavefunction

A universal wavefunction $|\Psi\rangle$ can be envisioned to live in $H_{\text{universal}}$. . It encapsulates all quantum states across the four layers, allowing for a holistic view of the system's evolution. Observable phenomena (e.g., space, time, energy, gravity) arise naturally in H_{L_c} but their origins are encoded in H_{L_∞}. If $|\psi_\infty\rangle \in H_{L_\infty}$, $|\psi_k\rangle \in H_{L_k}$, $|\psi_g\rangle \in H_g$, and $|\psi_c\rangle \in H_{L_c}$ represent quantum states in their respective Hilbert spaces, then the state in the universal Hilbert space is depicted by Eq. (3.2), *Universal Wavefunction*:

$$|\psi\rangle = |\psi_\infty\rangle \otimes |\psi_k\rangle \otimes |\psi_g\rangle \otimes |\psi_c\rangle. \tag{3.2}$$

Correspondingly, a universal operator, $\widehat{O_u}$, acting on $H_{\text{universal}}$, can be constructed from operators acting on individual spaces, depicted by Eq. (3.3), *Universal Operator*. Note that an operator imparts a certain action on a wavefunction within Hilbert spaces and is fundamental to extract information or even transform it, thereby rendering understanding, shaping, and even predicting behavior in any quantum space. In (3.3) $\widehat{O_\infty}$ acts on H_{L_∞}, $\widehat{O_k}$ acts on H_{L_k}, and so on:

$$\widehat{O_u} = \widehat{O_\infty} \otimes \widehat{O_k} \otimes \widehat{O_g} \otimes \widehat{O_c}. \tag{3.3}$$

This universal operator in (3.3) is illustrative of unifying transformations that simultaneously affect all the Hilbert spaces and is an upper-bound operator. Many other operators will be specified through the remainder of the book.

3.3 Quantum States Across Hilbert Space

At H_{L_∞}, where light is imagined traveling infinitely fast, the system retains the highest level of symmetry and abstraction. The quantum state $|\psi_\infty\rangle$ is expressed by Eq. (3.4), *The Quantum State at* H_{L_∞}:

$$|\psi_\infty\rangle = \sum_{i=1}^{N}\sum_{j=1}^{M}\sum_{k=1}^{P}\sum_{l=1}^{Q} c_{ijkl}|\text{Pr}_i\rangle \otimes \left|\text{Po}_j\right\rangle \otimes |K_k\rangle \otimes |H_l\rangle, \tag{3.4}$$

where:

- $|\text{Pr}_i\rangle$, $\left|\text{Po}_j\right\rangle$, $|K_k\rangle$, $|H_l\rangle$ represent the highest-level abstract states of Presence, Power, Knowledge, and Harmony.
- The coefficients c_{ijkl} encode relationships between these properties, potentially capturing perfect symmetry.

In H_{L_∞}, where light is envisioned traveling infinitely fast, spatial and temporal constraints are absent, allowing for maximal symmetry. As a result, c_{ijkl} may reflect symmetries related to high-dimensional groups like SU(n), SO(n) or U(n), because these groups [3] encode the transformations that preserve the structure and relationships between the quantum states. These coefficients determine how the basis states $|\text{Pr}_i\rangle$, $\left|\text{Po}_j\right\rangle$, $|K_k\rangle$, $|H_l\rangle$ combine to form the quantum state $|\psi_\infty\rangle$. Furthermore, c_{ijkl} inherently reflect the symmetries of the system, ensuring that the quantum state remains invariant under any transformations governed by groups such as SU(n), SO(n), $U(n)$, or any other.

The Hilbert space H_{L_k} encompasses all possible quantum states formed from combinations of elements in $S_{\text{Pr}}, S_{\text{Po}}, S_K, S_H$. These combinations could be interpreted as superpositions or as individual configurations of these elements. A quantum state $|\psi_k\rangle \in H_{L_k}$ is expressed by Eq. (3.5), *The Quantum State at* H_{L_k}:

$$|\psi_k\rangle = \sum_{i=1}^{R}\sum_{j=1}^{S}\sum_{k=1}^{T}\sum_{l=1}^{U} c_{ijkl}|\mathrm{Pr}_i\rangle \otimes |\mathrm{Po}_j\rangle \otimes |K_k\rangle \otimes |H_l\rangle, \tag{3.5}$$

where:

- $|\mathrm{Pr}_i\rangle$ represents a specific element $\mathrm{Pr}_i \in S_{\mathrm{Pr}}$.
- $|\mathrm{Po}_j\rangle$, $|K_k\rangle$, $|H_l\rangle$ similarly represent elements from their respective sets.
- $c_{ijkl} \in \mathbb{C}$ are the coefficients, representing the contribution of each combination to the state $|\psi_k\rangle$.

Each quantum state $|\psi_k\rangle$ is now a weighted combination of specific elements from $S_{\mathrm{Pr}}, S_{\mathrm{Po}}, S_K, S_H$, reflecting the richness and complexity of the derivative Hilbert space H_{L_k}.

At H_{L_g}, where seeds emerge as unique combinations of the sets in H_{L_k}, the quantum state $|\psi_g\rangle$ would focus on seed formation. It is expressed by Eq. (3.6), *The Quantum State at* H_{L_g}:

$$|\psi_g\rangle = \sum_{n=1}^{R} d_n|n\rangle, \tag{3.6}$$

where:

- $|n\rangle$ represents a specific seed formed by combinations of $|\mathrm{Pr}_i\rangle$, $|\mathrm{Po}_j\rangle$, $|K_k\rangle$, $|H_l\rangle$.
- d_n is the amplitude (or probability amplitude) associated with the seed $|n\rangle$
- Seeds $|n\rangle$ are derived from the function: $|n\rangle = f\left(|\mathrm{Pr}_i\rangle \otimes |\mathrm{Po}_j\rangle \otimes |K_k\rangle \otimes |H_l\rangle\right)$
- At this layer, the system transitions from structured sets to discrete seeds, emphasizing the emergence of unique quantum configurations.

At H_{L_c}, where light slows to the speed of c and space–time-energy-gravity properties manifest, the quantum state $|\psi_c\rangle$ represents quantum objects in physical spacetime. It can be expressed by Eq. (3.7), *The Quantum State at* H_{L_c}:

$$|\psi_c\rangle = \sum_{x=1}^{X}\sum_{t=1}^{T}\sum_{e=1}^{E}\sum_{g=1}^{G} \eta_{xteg}|S_x\rangle \otimes |T_t\rangle \otimes |E_e\rangle \otimes |G_g\rangle, \tag{3.7}$$

where:

- $|S_x\rangle$, $|T_t\rangle$, $|E_e\rangle$, $G_g\rangle$ represent basis states for space, time energy, gravity respectively.
- The coefficients η_{xteg} reflect the amplitudes that define the contributions for each specific space–time-energy-gravity configuration to the overall quantum state.

Seeds $|n\rangle$ project from H_{L_g} into H_{L_c} where they define specific space–time-energy-gravity configurations, governed by operators $\hat{S}, \hat{T}, \hat{E}, \hat{G}$. Physical properties (e.g., spatial location, temporal evolution) emerge as observable phenomena. This is summarized by Eqs. (3.8–3.11), *Operator Actions on Seeds*:

$$\widehat{S}|n\rangle = |S_x\rangle \tag{3.8}$$

$$\widehat{T}|n\rangle = |T_t\rangle \tag{3.9}$$

$$\widehat{E}|n\rangle = |E_e\rangle \tag{3.10}$$

$$\widehat{G}|n\rangle = \left|G_g\right\rangle. \tag{3.11}$$

Further, it is useful to illustrate mechanisms that project higher-dimensional wavefunctions onto lower layers by starting with $|\psi_\infty\rangle$ that contains all possible properties at the infinite-speed level. Such projection operators reduce the complexity of the state [4]. They impose deterministic constraints so that seeds encode certain properties while the resulting states retain probabilistic features.

A first projection, $\mathcal{P}_k$, results in $\mathcal{P}_k|\psi_\infty\rangle$ where the abstract properties Pr, Po, K, H organize into structured sets S_{Pr}, S_{Po}, S_K, S_H. A second, projection, $\mathcal{P}_g$, results in $\mathcal{P}_g(\mathcal{P}_k|\psi_\infty\rangle)$ where the seeds 'n' are formed as unique combinations of the sets S_{Pr}, S_{Po}, S_K, S_H. A third projection, $\mathcal{P}_c$, results in $\mathcal{P}_c\left(\mathcal{P}_g(\mathcal{P}_k|\psi_\infty\rangle)\right)$ where the seeds manifest as quantum objects with specific space–time–energy–gravity properties. This is summarized by Eq. (3.12), *Projection of Higher Dimensional Wavefunctions onto Lower Layers*:

$$|\psi_c\rangle = \mathcal{P}_c\mathcal{P}_g\mathcal{P}_k|\psi_\infty\rangle. \tag{3.12}$$

3.4 Eigenstates and Eigenvalues

An eigenstate [5] represents the possible state of a system that yields a definite, predictable value when a measurement of a particular observable is performed. An eigenvalue represents the specific, quantized value obtained when measuring the corresponding observable. At H_{L_∞}, eigenvalues and eigenstates reflect the most abstract and undifferentiated form of the system. Operators such as $\widehat{\mathrm{Pr}_\infty}$ (Presence), $\widehat{\mathrm{Po}_\infty}$ (Power), $\widehat{K_\infty}$ (Knowledge), and $\widehat{H_\infty}$ (Harmony) act on the basis states and the eigenvalues p_i, q_j, r_k, s_l at this layer reflect high-dimensional properties that are invariant under any symmetry transformations (e.g., SU(n), SO(n), $U(n)$) and are depicted by Eqs. (3.13)–(3.16), *H_{L_∞} Eigenstates and Eigenvalues*:

$$\widehat{\mathrm{Pr}_\infty}\left|\mathrm{Pr}_{\infty,i}\right\rangle = p_i\left|\mathrm{Pr}_{\infty,i}\right\rangle \tag{3.13}$$

$$\widehat{\mathrm{Po}_\infty}\left|\mathrm{Po}_{\infty,j}\right\rangle = q_j\left|\mathrm{Po}_{\infty,j}\right\rangle \tag{3.14}$$

$$\widehat{K_\infty}\left|K_{\infty,k}\right\rangle = r_k\left|K_{\infty,k}\right\rangle \tag{3.15}$$

$$\widehat{H_\infty}\left|H_{\infty,l}\right\rangle = \mathrm{s}_l\left|H_{\infty,l}\right\rangle. \tag{3.16}$$

As symmetry breaks at H_{L_k}, projection operators project the quantum state $|\psi_\infty\rangle$ into structured subsets S_{Pr}, S_{Po}, S_K, S_H. Eigenvalues at this layer define deterministic properties of subsets where $\lambda_{\mathrm{Pr},i}$ captures structured properties of Presence, Power, Knowledge, Harmony, as depicted by Eqs. (3.17–3.20), H_{L_k} *Eigenstates and Eigenvalues:*

$$\widehat{\mathrm{Pr}_k}\left|\mathrm{Pr}_{k,i}\right\rangle = \lambda_{\mathrm{Pr},i}\left|\mathrm{Pr}_{k,i}\right\rangle \tag{3.17}$$

$$\widehat{\mathrm{Po}_k}\left|\mathrm{Po}_{k,j}\right\rangle = \lambda_{\mathrm{Po},j}\left|\mathrm{Po}_{k,j}\right\rangle \tag{3.18}$$

$$\widehat{K_k}\left|K_{k,k}\right\rangle = \lambda_{K,k}\left|K_{k,k}\right\rangle \tag{3.19}$$

$$\widehat{H_k}\left|H_{k,l}\right\rangle = \lambda_{H,l}\left|H_{k,l}\right\rangle. \tag{3.20}$$

At H_{L_g}, seeds $|n\rangle$ emerge as eigenstates of a composite seed operator where f is a function defining how the fundamental properties (Presence, Power, Knowledge, Harmony) combine. The composite seed operator is depicted by Eq. (3.21), H_{L_g} *Composite Seed Operator*, where the subscript 'g' refers to this Hilbert space conceptualized when light travels at constant speed 'g':

$$\widehat{O_{s@g}} = f\left(\widehat{\mathrm{Pr}_g}, \widehat{\mathrm{Po}_g}, \widehat{K_g}, \widehat{H_g}\right). \tag{3.21}$$

The eigenvalues λ_n satisfying an eigenvalue equation for the operator $\widehat{O_{s@g}}$ as depicted by Eq. (3.22), *Eigenvalue Equation for Operator* $\widehat{O_{s@g}}$:

$$\widehat{O_{s@g}}|n\rangle = \lambda_n|n\rangle. \tag{3.22}$$

At H_{L_c}, seeds project into quantum objects where eigenvalues represent values related to the specific space–time-energy-gravity code that animates a quantum object. Note that the subscript 'c' refers to this Hilbert space conceptualized when light travels at constant speed 'c':

- Knowledge (x): Eigenvalues of the knowledge operator $\widehat{K_c}$, depicted by 'x' to continue to suggest that all space is granular and populated by seeds that have meaning.

- Power (t): Eigenvalues of the power operator $\widehat{\mathrm{Po}_c}$, depicted by 't' to continue to suggest that seeds exercise power to reveal what they mean, through the passage of time.
- Presence (e): Eigenvalues of the energy operator $\widehat{\mathrm{Pr}_c}$, depicted by 'e' to continue to suggest that seeds transform from subtle to material by conversion of energy.
- Harmony (g): Eigenvalues of the gravitational operator $\widehat{H_c}$, depicted by 'g' to continue to suggest that seeds are always in some collectivity or harmony with other select seeds.

Corresponding eigenstates and eigenvalues are depicted by Eqs. (3.23–3.26), H_{L_c} *Eigenstates and Eigenvalues*:

$$\widehat{K_c}|n\rangle = x_n|n\rangle \tag{3.23}$$

$$\widehat{\mathrm{Po}_c}|n\rangle = t_n|n\rangle \tag{3.24}$$

$$\widehat{\mathrm{Pr}_c}|n\rangle = e_n|n\rangle \tag{3.25}$$

$$\widehat{H_c}|n\rangle = g_n|n\rangle. \tag{3.26}$$

3.5 Determinism and the Role of Seeds

The seeds in H_{L_g} are not merely probabilistic configurations. They act as blueprints for quantum objects that surface at H_{L_c}. Each seed $n \in H_{L_g}$, derived from $f(S_{\mathrm{Pr}}xS_{\mathrm{Po}}xS_K xS_H)$, uniquely encodes the specific space–time-energy-gravity profile for a quantum object. The seed acts as a functional mapping that takes the rich information from the sets S_{Pr}, S_{Po}, S_K, S_H and encodes it into a unique combination. No two seeds encode the exact same space–time–energy–gravity profile. This deterministic aspect ensures that any quantum object that emerges at H_{L_c} has a defined structure and behavior, rooted in the properties of its seed. While probabilities in quantum mechanics describe how a seed interacts with the environment or external conditions, from this point of view, the seed itself imposes constraints on what is possible.

By definition, all quantum objects arising in H_{L_c} inherently have space, time, energy, and gravity as embedded components, where 'space' specifies how the object will behave, 'time' specifies the object's temporal evolution, 'energy' encodes the object's transformation profile, and 'gravity' encodes the object's relationship with other objects through curvature or mutual influence. Thus, every quantum object can be thought of as a four-dimensional quantum unit, where these dimensions are inseparable and emerge as a natural consequence of the seed's structure. That is, each seed 'n' represents a distinct quantum state that encodes aspects of Presence, Power, Knowledge, and Harmony in a way that allows for the emergence of space–time–energy–gravity relationships.

At H_{L_g}, seeds are still quantum configurations. However, when they manifest in H_{L_c}, they define objects that behave with greater determinism, bridging the quantum and classical worlds. The deterministic nature of seeds helps explain why macroscopic phenomena emerge as structured systems, even though they originate from quantum properties.

Equations (3.13–3.26), hence, refer to operators, eigenstates, and eigenvalues related to the formation of seeds.

It is also important to understand how seeds evolve dynamically. That is, what governs the evolution of seeds in H_{L_g}? To understand this, it is necessary to define a Hamiltonian $\widehat{H_g}$ that governs the seeds dynamics, depicted by Eq. (3.27), *Hamiltonian that Governs Seed Dynamics*:

$$i\hbar\frac{\partial|n(t)\rangle}{\partial t} = \widehat{H_g}|n(t)\rangle. \tag{3.27}$$

This implies a system of feedback between layers that will be addressed in a later chapter. Hamiltonians are discussed in greater detail in Chap. 7 and (3.27) surfaces it in the context of the possible changing dynamics of seeds. This equation is a time-dependent Schrödinger equation, which describes how a quantum state $|n(t)\rangle$ evolves over time. At a high-level:

1. $|n(t)\rangle$ is the quantum state of the system at time t.
2. $\widehat{H_g}$ is the Hamiltonian—the energy operator that governs how the system evolves.
3. Equation (3.27) says that the rate of change of the quantum state is directly determined by applying the Hamiltonian to that state.

Putting it simply, the Hamiltonian "tells" the quantum state how to change in time, and the Schrödinger equation [6] is the core rule in quantum mechanics that ties a system's energy properties (the Hamiltonian) to the way its quantum state behaves from one moment to the next.

3.5.1 Probabilistic and Deterministic Balance

The density matrix plays an important role in the emergence and functioning of seeds as it can capture both probabilistic and deterministic aspects of a quantum system. The states $|n\rangle$ encode the deterministic structure of the seeds. The probabilities p_n reflect the uncertainty in the state of the system. But further, mixed states allow multiple seeds $|n\rangle$ to coexist with different probabilities p_n. This reflects how a quantum object can emerge from a probabilistic combination of deterministic seed properties. In Eq. (3.28), *The Density Matrix*, the density matrix is represented by ρ:

$$\rho = \sum_n p_n |n\rangle\langle n|. \tag{3.28}$$

Here:

- ρ: The density operator encodes all the statistical information about a quantum system's state, including both pure and mixed states.
- $|n\rangle$: Each $|n\rangle$ is a quantum state—one of a set of basis states.
- $|n\rangle\langle n|$: This is the projector onto the state $|n\rangle$. It can be thought of as an operator that filters the component of any vector along $|n\rangle$ and is further elaborated by (3.32) below.
- p_n: A real, non-negative number representing the probability associated with $|n\rangle$. All p_n sum to 1.

When there are multiple seeds, each $|n\rangle$ contributes deterministically to the system's structure. The probabilities p_n describe how likely the system is to adopt each seed's properties.

To measure an observable property $\hat{O}$, the expectation value can be computed with the trace operator, Tr, where ρ is the density matrix and incorporates both probabilistic outcomes and deterministic seed structure. Note that the trace function, Tr, ensures that the density matrix is normalized by summing the probabilities encoded in its diagonal elements [7]. This guarantees that the total probability of all quantum states is 1. The measured value $\langle\hat{O}\rangle$ reflects contributions from all seeds $|n\rangle$, weighted by their probabilities p_n, and is depicted by Eq. (3.29), *Expectation Value of Observable Property*:

$$\langle\hat{O}\rangle = \mathrm{Tr}(\rho\hat{O}). \tag{3.29}$$

If, for example, $\hat{S}$ is the operator for spatial dimension, the expected value of x, which combines the deterministic contribution from each seed $|n\rangle$ (via $\langle n|\hat{S}|n\rangle$) with the probabilistic weighting p_n is depicted by Eq. (3.30), *Expected Value of x*:

$$\langle x\rangle = \mathrm{Tr}(\rho\hat{S}) = \sum_n p_n\langle n|\hat{S}|n\rangle. \tag{3.30}$$

In the case where a quantum system is pure, that is, is definitely in the state $|n\rangle$ with no ambiguity or probability of being in any other state, the density matrix ρ is depicted by Eq. (3.31), *Pure State Density Matrix*:

$$\rho = |n\rangle\langle n|. \tag{3.31}$$

The way to think about this is as a projection operator onto the quantum state $|n\rangle$. That is, it acts on a quantum state $|\psi\rangle$ to project $|\psi\rangle$ onto the subspace spanned by $|n\rangle$ or to extract the part of $|\psi\rangle$ that lies in the subspace spanned by $|n\rangle$, as depicted by Eq. (3.32), *Extraction of Quantum State*:

$$(|n\rangle\langle n|)|\psi\rangle = \langle n|\psi|n\rangle. \tag{3.32}$$

The result is the component $|\psi\rangle$ along $|n\rangle$, scaled by the overlap $\langle n|\psi\rangle$.

3.6 Further Reading

- **"Symmetry" by Hermann Weyl:** This is a classic and highly influential book that explores the concept of symmetry in mathematics, science, and art. It provides a foundational understanding of symmetry and its importance in various fields.
- **"Broken Symmetry" by Philip W. Anderson:** This book delves into the concept of symmetry breaking in condensed matter physics and its role in emergent phenomena like superconductivity and magnetism.
- "More Is Different" by Philip W. Anderson: This influential paper argues that emergent phenomena in complex systems cannot be simply reduced to the properties of their individual components. It emphasizes the importance of interactions and collective behavior in giving rise to new properties at macroscopic scales.
- "Methods of Mathematical Physics, Vol. 1" by Richard Courant and David Hilbert: This classic text provides a rigorous mathematical foundation for quantum mechanics, including a detailed treatment of Hilbert spaces and linear operators.
- "Group Theory in Physics" by Wu-Ki Tung: This book provides a comprehensive introduction to group theory with a focus on its applications in physics. It includes detailed discussions of Lie groups, including SU(n), SO(n), and U(n).
- "Functional Analysis" by Peter Lax: This book offers a rigorous treatment of functional analysis, including a detailed discussion of projections in Hilbert spaces. This is a more advanced text suitable for those with a strong mathematical background.
- "Modern Quantum Mechanics" by J. J. Sakurai: Sakurai discusses the trace in the context of density matrices, which are essential for describing mixed states in quantum mechanics. The trace is used to calculate expectation values of observables in mixed states.
- "A Brief History of String Theory: From Dual Models to M-Theory" by Dean Rickles: This book provides a historical perspective on the development of string theory, making it more approachable for those interested in the evolution of scientific ideas.

References

1. Kadison RV (1968) Tensor product of Hilbert spaces. University of Texas at Austin, Austin
2. Zhou F (2024) Emergent symmetries and interactions in high dimensions: an isolated fixed point versus a manifold of strongly interacting fixed points. Phys Rev 109(18): Art. no. 184503, APS. https://doi.org/10.1103/PhysRevB.109.184503

3. Hall BC (2015) Lie groups, lie algebras, and representations: an elementary introduction, 2nd edn. Springer, New York
4. Courant R, Hilbert D (1953) Methods of mathematical physics, vol 1. Interscience Publishers, New York
5. Griffiths DJ (2004) Introduction to quantum mechanics, 2nd edn. Pearson Prentice Hall, Upper Saddle River, NJ
6. Schrödinger E (1926) An undulatory theory of the mechanics of atoms and molecules. Phys Rev 28(6):1049–1070. https://doi.org/10.1103/PhysRev.28.1049
7. Hoffman K, Kunze R (1971) Linear algebra, 2nd edn. Prentice-Hall, Englewood Cliffs, NJ

Chapter 4
The Mathematics of Qualified Determinism

Abstract This chapter proposes a novel framework for quantum computation based on "qualified determinism" (QD) and a multi-layered model of quantum objects, challenging the conventional focus on probabilistic interpretations. The "Qualified Determinism Construct" (QDC) posits that quantum objects possess underlying deterministic blueprints accessed through meta-layers ($M1$, $M2$, $M3$) interacting with the conventional "base layer" (B-Layer). Each meta-layer introduces distinct dynamics (o, n, m-entanglement) influencing the B-Layer's qubits (termed "qibits"), gates, and algorithms. Computations leveraging these meta-layers ($B + Mx$) are termed "l-stratum," dealing with "wholes," contrasting with the B-Layer's "r-stratum" probabilistic computations. The framework suggests that the level of determinism and superposition dynamics change based on active meta-layers, even persisting during B-Layer measurements. Finally, the concept of "X," a foundational "quantum-functional seed" embodying "Organizing Quantum Energy" (OQE) as both informational blueprint and driving force for deterministic organization, is introduced. This framework aims to shift quantum computation toward deterministic interpretations through novel architectures, hardware, and algorithms.

The previous chapter introduced various structures and mechanisms tied to the evolution of quantum systems. These artifacts are all founded on the notion that to harness quantum technology aligned with $u(W(L))$ and $v(W(L))$ type quantum computation, which deals in 'wholes', it is necessary to reframe quantum objects as a balance between deterministic structure and quantum uncertainty. Such quantum objects can be thought of as subject to 'qualified determinism' (QD).[1] Note that Bohm postulated that randomness observed in experiments does not come from the nature of reality itself, but rather from our ignorance of hidden variables [1, 2]. Further, his postulation supported nonlocality in line with implications of entanglement and Bell's theorem [3]. Operators, eigenstates, and eigenvalues then become part of the parsing mechanism to tease out such wholeness, which perhaps can be thought of as a form

[1] An early non-quantum version of QD appeared in an IEEE article that has been referenced in the Further Reading section of this chapter.

P. Malik, *Pioneering New Avenues in Quantum Technology*, Studies in Smart Technologies, https://doi.org/10.1007/978-981-96-5463-5_4

of nonlocality, from different species of wavefunctions that live in corresponding Hilbert spaces.

This chapter will build on the previous chapter to explore mathematics related to QD-based quantum objects that will be instrumental in thinking about quantum technology differently.

4.1 A Universal Quantum Object

To begin with a Universal Quantum Object (UQO) will be created by combining many of the equations constructed in Chap. 3. Hence, Eqs. (4.1–4.5) are created leveraging (3.4–3.7).

Equation (4.1), Summary of Quantum State at H_{L_∞}, creates a summary 'm' function from (3.4):

$$|\psi_\infty\rangle = m(\text{Pr}, \text{Po}, K, H) = \sum_{i=1}^{N}\sum_{j=1}^{M}\sum_{k=1}^{P}\sum_{l=1}^{Q} c_{ijkl}|\text{Pr}_i\rangle \otimes |\text{Po}_j\rangle \otimes |K_k\rangle \otimes |H_l\rangle. \quad (4.1)$$

Equation (4.2), Summary of Quantum State at H_{L_k}, creates a summary 'n' function from (3.5):

$$\begin{aligned} |\psi_k\rangle &= n(S_{\text{Pr}}, S_{\text{Po}}, S_K, S_H) \\ &= \sum_{\text{Pr}_i \in S_{\text{Pr}}} \sum_{\text{Po}_j \in S_{\text{Po}}} \sum_{K_k \in S_K} \sum_{H_l \in S_H} c_{ijkl}|\text{Pr}_i\rangle \otimes |\text{Po}_j\rangle \otimes |K_k\rangle \otimes |H_l\rangle \end{aligned} \quad (4.2)$$

Equation (4.3), Summary of Quantum State at H_{L_g}, creates a summary 'o' function from (3.6):

$$|\psi_g\rangle = o(S_{\text{Pr}} \times S_{\text{Po}} \times S_K \times S_H) = \sum_{n=1}^{R} d_n|n\rangle. \quad (4.3)$$

Equation (4.4), Summary of Quantum State at H_{L_c}, creates a summary 'p' function from (3.7):

$$|\psi_c\rangle = p(S, T, E, G) = \sum_{x=1}^{X}\sum_{t=1}^{T}\sum_{e=1}^{E}\sum_{g=1}^{G} \eta_{xteg}|S_x\rangle \otimes |T_t\rangle \otimes |E_e\rangle \otimes |G_g\rangle. \quad (4.4)$$

Equation (4.5), Universal Quantum Object, combines (4.1–4.4) with the projection functions P_k, P_g, and P_c referred to in (3.12):

$$\text{UQO} = \begin{bmatrix} |\psi_\infty\rangle = m(\text{Pr}, \text{Po}, K, H) \\ (\downarrow P_k|\psi_\infty\rangle) \\ |\psi_k\rangle = n(S_{\text{Pr}}, S_{\text{Po}}, S_K, S_H) \\ (\downarrow P_g|\psi_k\rangle) \\ |\psi_g\rangle = o(S_{\text{Pr}} \times S_{\text{Po}} \times S_K \times S_H) \\ (\downarrow P_c|\psi_g\rangle) \\ |\psi_c\rangle = p(S, T, E, G) \end{bmatrix}. \tag{4.5}$$

In (4.5), the structure of a UQO is such that there are many layers that contribute to it. Yet in our approach to quantum computation, we appear to be focused on $|\psi_c\rangle$ only, and being so, approach the computation by relying on probability and statistics, as suggested in the discussion on the r-stratum in Chap. 1.

4.2 A Qualified Determinism Construct

The issue is what can we do differently to begin to perceive and even compute as per the blueprint set up by the antecedent seed referred to by the '*o*' function, and perhaps even as per pure property referred to by the '*n*' function, or even the complex agglomerate referred to by the '*m*' function in (4.5)? Note that Chaps. 13 through 16 illustrate the application of the *o*, *n*, *m* functions or operators in creating the structure of the electromagnetic spectrum, quantum particles, atoms, and molecular plans in cells, respectively. Here we can derive a construct from (4.5)—a Qualified Determinism Construct (QDC)—that can potentially be used as an aid to assess the kind of computation we are involved in.

If computation proceeds as per the '*p*' function only, then as per (4.4), the eigenstates and eigenvalues as suggested by (3.23–3.26) will appear as surface observables of the type x_n, t_n, e_n, and g_n. One would than repeat the quantum calculation many times, and through statistics and probability, infer a single value for the calculation. The point though is that the deeper deterministic blueprint that deals with the *l*-stratum, and therefore with $u(W(L))$ or $v(W(L))$ wholes, is not being leveraged. Algorithms and the hardware set up a blueprint with variables, constraints, and mechanisms to utilize these—but it is all taking place in the *r*-stratum only. Leveraging (4.4), we can simply state this by Eq. (4.6), *Base Layer Operations*, where '*B*' refers to the base layer:

$$B \rightarrow |\psi_c\rangle = p(S, T, E, G). \tag{4.6}$$

Entanglement is created and will be referred to as Entanglement$_p$ reinforcing its link with the '*p*' function, and depicted by Eq. (4.7), *P*-Entanglement:

$$\text{Entanglement}_p \propto p(S, T, E, G). \tag{4.7}$$

If computation proceeds as per the 'o' function, then as per (4.3), the eigenstates and eigenvalues as suggested by (3.21–3.22) will yield a configuration for a seed, depicted by Eq. (4.8), *Configuration for Seed*:

$$\text{Seed} = Xa + \sum_{i=0}^{n} Y_i b_i \text{ where} \begin{bmatrix} X, Y \in [S_{\text{Pr}}, S_{\text{Po}}, S_K, S_H] \\ a, b \text{ are integers; } a > b \end{bmatrix}. \tag{4.8}$$

Such seeds are hypothesized to be instrumental in the surfacing of functionality that accompanies constructs in different layers of matter—quantum particles, atoms, molecules, and the layer of life—molecular plans in cells. Leveraging (4.3), we can state this by Eq. (4.9), *$M1$ Operations*, where $M1$ refers to meta-layer 1:

$$M1 \rightarrow |\psi_g\rangle = o(S_{\text{Pr}} \times S_{\text{Po}} \times S_K \times S_H). \tag{4.9}$$

Computation that leverages an $M1$-dynamic could hypothetically affect the nature of a seed or could be utilized in creating or imprinting its properties in a target qibit (quaternary interpretation-based 'qubit') that would be subject to the computation. Further, the $M1$ dynamic suggests an o-entanglement, tied to the o-function, by which uniqueness of seeds is ensured. This type of entanglement is therefore already assumed to exist, and the M1-dynamic can be thought of as manipulating it. This kind of entanglement, Entanglement$_O$, is depicted by Eq. (4.10), O-Entanglement:

$$\text{Entanglement}_O \propto o(S_{\text{Pr}} \times \text{S}_{\text{Po}} \text{ x } \text{S}_{\text{K}} \text{ x } \text{S}_{\text{H}}). \tag{4.10}$$

If computation proceeds as per the 'n' function, then as per (4.2) the eigenstates and eigenvalues as suggested by (3.17–3.20) will yield a pure property derived from an element from one of the four sets, depicted by Eq. (4.11), $M2$ Operations:

$$M2 \rightarrow |\psi_k\rangle = n(S_{\text{Pr}}, S_{\text{Po}}, S_K, S_H). \tag{4.11}$$

If in a computational architecture, then, if seeds or qibits were subject to an $M2$-dynamic, it could hypothetically affect that specific property in the configuration of a seed or imprint its essence in the target qibit. The $M2$ dynamic suggests an n-entanglement, tied to the n-function, by which 'purity' affect the creation or modification of seeds or qibits that have that property as part of their configuration. This is summarized by Eq. (4.12), *N-Entanglement*, where Entanglement$_n$ depicts this kind of entanglement:

$$\text{Entanglement}_n \propto n(S_{\text{Pr}}, S_{\text{Po}}, S_K, S_H). \tag{4.12}$$

If computation proceeds as per the 'm' function, then as per (4.1), the eigenstates and eigenvalues as suggested by (3.13–3.16) will yield a complex agglomerate, depicted by Eq. (4.13), *$M3$ Operations*:

$$M3 \rightarrow |\psi_{\infty}\rangle = m(\text{Pr}, \text{Po}, K, H). \tag{4.13}$$

A living cell with its range of molecular plans, and trillions of atoms, can be thought of as an example of a 'complex agglomerate'. As per the quantum model being developed in this book, thousands upon thousands of properties manifest by combining with material form, and there is an integrated or unified fourfold complexity—a complex agglomerate—that drives the functioning of the overall cell. Recent research suggests that several quantum phenomena have been observed or theorized to occur within cells, potentially impacted by the cohesive environment that a cell provides [4–7].

If in a computational architecture, then, if seeds or qibits were subject to an $M3$-dynamic, it could hypothetically affect that functioning of a target agglomerate or qibit. The $M3$-dynamic suggests an m-entanglement, tied to the m-function, by which fourfold complexity as manifest in a source cell would affect the creation or modification of target qibits as per its signature of fourfold complexity. The action of entanglement and the power of superposition are hypothesized as being at a higher level that the just discussed dynamics or types of entanglement. The m-entanglement is summarized by Eq. (4.14), *M-Entanglement*:

$$\text{Entanglement}_m \propto m(\text{Pr}, \text{Po}, K, H). \tag{4.14}$$

A Qualified Determinism Construct, QDC, is constructed by arranging (4.6, 4.9, 4.11, 4.13) in the matrix form specified by Eq. (4.15), *Qualified Determinism Construct*:

$$\text{QDC} = \begin{bmatrix} M3 \rightarrow m(\text{Pr}, \text{Po}, K, H) \\ \downarrow\uparrow \\ M2 \rightarrow n(S_{\text{Pr}}, S_{\text{Po}}, S_K, S_H) \\ \downarrow\uparrow \\ M1 \rightarrow o(S_{\text{Pr}} \times S_{\text{Po}} \times S_K \times S_H) \\ \downarrow\uparrow \\ B \rightarrow p(S, T, E, G) \end{bmatrix}. \tag{4.15}$$

4.3 Some Quantum Computing Implications of QDC

QDC connects together different causal layers.

The B-Layer is driven by the quantum mechanical properties of entanglement and superposition, as commonly interpreted. Sense-making is driven by probability and statistics founded on a paradigm of measurement that keeps the focus on surface variables. In effect, the qubits, gates, and algorithms are built to reinforce an r-stratum type computational stratum, referred to as the 'base' of B-Layer in the QDC.

Quantum computation can proceed differently though when the context of dynamics that occurs in the *B*-Layer is seen to be set by any of the meta-layers—$M1$, $M2$, or $M3$. This will have a profound effect on the type of hardware and algorithms that will be required and will be taken up in more detail in Chaps. 18 and 19. The hardware and algorithms will be based on a Quaternary Interpretation of Quantum Dynamics (QIQD), to be summarized in Chap. 17, and the 'qubit' as suggested earlier will be a quaternary interpretation (QI) type quantum bit—qibit.

A primarily leveraged computational object at $M1$ will be atoms. These atoms, the carriers or vehicles of specific sets of properties dictated by the configuration of their seeds, will be part of the hardware and will act on qibits via gates and algorithms to impart the output of the computation in them. The output may not just be a representation of a concept, but an actual quality created by a property or set of properties that will live in that qibit. Gates will be constructed by different arrangements of atoms that energetically will 'compute' such qualities.

At $M2$ too, the primary computational object is envisioned to be atoms arranged in such a way so as to impart pure property to qibits, and selected portions of the electromagnetic spectrum that would be easier to manipulate and separate by underlying property. At $M3$, the primary computational object will be complex qibit agglomerates even approaching some of the functionality that exists in cells. These are seen as carriers of 'complex agglomerates' that also imprint another type of quality into target qibits or a range of qibits.

The qibit can also be thought of as being a new hardware entity at the *B*-Layer which is loaded with a particular kind of deterministic structure. The inclusion of this at the *B*-Layer will necessitate different paradigms of gates and hardware, and algorithms. Further, a *B*-Layer qibit may have counterparts at $M1$, $M2$, and / or $M3$ and is something that will need to be worked out by advancement of quantum technology. Equation (4.16), *Variability in Structure of Qibit*, summarizes the variability that may exist in the structure of the qibit:

$$qibit = f(\text{QDC}). \tag{4.16}$$

To summarize, the quantum computational architectures will be different for different combinations of layers in QDC. The architecture for '*B* type' computers will be fundamentally different than a '$B + Mx$ type' computer, where x can refer to any or all of the meta-layers. As suggested, while the former adheres to the notion of *r*-stratum, the latter can be thought of as an *l*-stratum quantum computer, where it is incorporating logic based on wholes.

Equation (4.17), *B-Layer Entanglement*, summarizes the type of entanglement that the *B*-Layer may be subject to, depending on the type of qibit created:

$$B\text{ - Layer Entanglement} = f\begin{pmatrix} \text{Entanglement}_m\text{:}M3 \rightarrow \text{complex agglomerate} \\ \downarrow\uparrow \\ \text{Entanglement}_n\text{:}M2 \rightarrow \text{pure property} \\ \downarrow\uparrow \\ \text{Entanglement}_o\text{:}M1 \rightarrow \text{set of properties} \\ \downarrow\uparrow \\ \text{Entanglement}_p\text{:}B \rightarrow \text{constructed} \end{pmatrix}. \tag{4.17}$$

4.4 Determinism and Quantum Utility

The level of determinism is going to depend on the type of qibit used as the basis of a quantum computing architecture. This qibit will be subject to different types of entanglement as suggested by (4.17), where the level of determinism will increase if any of the meta-layers are active. But further, the type of superposition will also change based on the qibit influenced by the active layers. Even if a measurement takes place in the B-Layer and $|\psi_c\rangle$ collapses, yet $|\psi_g\rangle$, $|\psi_k\rangle$, and $|\psi_\infty\rangle$ remain active allowing the effects of superposition to persist. This continuity becomes clearer when computational objects from $M1$, $M2$, and / or $M3$ are explicitly incorporated in the quantum computational architecture.

This concept is formally expressed by Eq. (4.18), Quantum Utility:

$$\text{Quantum Utility} = \begin{bmatrix} M3 \rightarrow m(\text{Pr}, \text{Po}, K, H) \\ \downarrow\uparrow \\ M2 \rightarrow n(S_{\text{Pr}}, S_{\text{Po}}, S_K, S_H) \\ \downarrow\uparrow \\ M1 \rightarrow o(S_{\text{Pr}} \times S_{\text{Po}} \times S_K \times S_H) \\ \downarrow\uparrow \\ B \rightarrow p(S, T, E, G) \end{bmatrix} \rightarrow C(\rho_{M_3}) > C(\rho_{M_2}) > C(\rho_{M_1}) > C(\rho_B), \tag{4.18}$$

where $C(\rho) = \sum_{i \neq j} |\rho_{ij}|$, C is the coherence [8] of the quantum state, and ρ is the density matrix with $i \neq j$ setting the focus to non-diagonal or a mixed quantum state. The higher C is, the more superposed the quantum system is. This has implications for quantum computation in that the likelihood of the system exploring a larger part of Hilbert space increases, which enables quantum parallelism and increased computational power. In quantum sensing higher superposition allows for greater sensitivity to changes in environmental parameters. Higher coherence also generally correlates with greater possibility of entanglement.

4.5 UQO, X, and OQE

In the framework being created here, X is introduced as the foundational quantum-functional seed, a unified archetype of information and energy. Unlike conventional energy, which arises from interactions, X inherently embodies 'Organizing Quantum Energy' (OQE), the intrinsic force driving deterministic organization across all scales. X, therefore, manifests as both the informational blueprint encoding universal properties—Knowledge, Presence, Harmony, and Power—and the energy (OQE) necessary to transform and propagate these properties.

The Universal Quantum Object (UQO) framework governs the transformation of X across layers of organization. Equation (4.19) formalizes this relationship, showing that the process begins with the m-operator, which acts first to manifest X as emergent structures energized by OQE. These structures are further refined by the n-operator, which integrates them into pure property sets, and by the o-operator, which configures these sets into seeds that propagate deterministic organization. This sequence ensures that X is realized across levels, from quantum particles to macroscopic systems.

At each level, X and OQE drive the emergence of structure. At the electromagnetic spectrum level, X structures the interplay of wavelength, frequency, and energy, shaping light and spacetime dynamics (discussed in Chap. 13). At the quantum particle level, the m-operator manifests quarks, leptons, bosons, and the Higgs boson, each representing one of the fourfold properties (discussed in Chap. 14). At the atomic level, X organizes elements into p-, d-, s-, and f-Groups, defining chemical behaviors (discussed in Chap. 15). At the macromolecular level, X governs nucleic acids as knowledge-encoding entities, proteins as embodiments of presence, lipids as harmonizing agents, and polysaccharides as drivers of power (discussed in Chap. 16).

UQO (4.5) is refined to include X—the unified information-energy archetype, as depicted by Eq. (4.19), UQO *(X)*, captures this layered progression:

$$\mathrm{UQO}(X) = \begin{bmatrix} |\psi_\infty\rangle = m(X; \mathrm{Pr}, \mathrm{Po}, K, H) \\ (\downarrow P_k|\psi_\infty\rangle) \\ |\psi_k\rangle = n(X; S_{\mathrm{Pr}}, S_{\mathrm{Po}}, S_K, S_H) \\ (\downarrow P_g|\psi_k\rangle) \\ |\psi_g\rangle = o(X; S_{\mathrm{Pr}} \times S_{\mathrm{Po}} \times S_K \times S_H) \\ (\downarrow P_c|\psi_g\rangle) \\ |\psi_c\rangle = p(X; S, T, E, G) \end{bmatrix}. \tag{4.19}$$

The concept of X will be instrumental to the discussion in Chaps. 13, 14, 15, and 16, that delve into the Quaternary Interpretation of Quantum Dynamics (QIQD) summarized in Chap. 12.

4.6 Further Reading

- "The Undivided Universe: An Ontological Interpretation of Quantum Theory" by David Bohm and B. J. Hiley: Bohmian mechanics offers a deterministic interpretation of quantum mechanics by introducing hidden variables that determine the trajectories of particles. This is a controversial but important perspective.
- "Decoherence and the Quantum-to-Classical Transition" by Maximilian Schlosshauer: Decoherence theory explains how interactions with the environment lead to the suppression of quantum superpositions and the emergence of classical behavior. This helps to understand why we observe determinism at the macroscopic level even though the underlying quantum world is probabilistic.
- "Natural Philosophy of Cause and Chance" by Max Born: Born emphasized the probabilistic nature of quantum mechanics, arguing that quantum mechanics abandoned determinism in favor of probabilistic predictions.
- "Quantum Measurement" by Vladimir B. Braginsky and Farid Ya. Khalili: This book provides a detailed and rigorous treatment of quantum measurement theory, which is crucial for understanding how measurements affect quantum systems and the apparent transition from quantum uncertainty to classical determinism.
- "Speakable and Unspeakable in Quantum Mechanics" by John S. Bell: A collection of essays that critically examines the foundations and interpretations of quantum theory, particularly focusing on the implications of quantum nonlocality. Bell's work challenges the prevailing Copenhagen interpretation and its avoidance of clear ontological statements about the nature of reality.
- "Life on the Edge: The Coming of Age of Quantum Biology" by Johnjoe McFadden and Jim Al-Khalili: This book provides an accessible introduction to quantum biology, explaining how quantum phenomena like superposition and tunneling may play a role in processes like photosynthesis, bird migration, and even consciousness.
- "Dancing with Qubits" by Robert S. Sutor: This book provides an intuitive and less mathematically rigorous introduction to quantum computing.
- "Deterministic Quantum Mechanics: The Mathematical Equations" by Gerard 't Hooft: Presents a mathematical framework for describing quantum systems in a deterministic manner. 't Hooft argues that quantum mechanics can be mathematically equivalent to deterministic systems under certain conditions. He aims to write down the equations that make this equivalence possible without delving into philosophical implications.
- "Qualified determinism in emergent-technology Complex Adaptive Systems" by Malik, P. et al.: An early version of the idea of qualified determinism as applied to complex behavioral complex systems that is a precursor for the mathematics developed here.

References

1. Bohm D (1952) A suggested interpretation of the quantum theory in terms of hidden variables I & II. Phys Rev 85(2):166–193. https://doi.org/10.1103/PhysRev.85.166
2. Holland PR (1993) The quantum theory of motion: an account of the de Broglie–Bohm causal interpretation of quantum mechanics. Cambridge University Press
3. Bell JS (1964) On the Einstein Podolsky Rosen paradox. Physics 1(3):195–200
4. Raković D, Dugić M, Jeknić-Dugić J, Plavšić M, Jaćimovski S, Setrajčić J (2014) On macroscopic quantum phenomena in biomolecules and cells: from Levinthal to Hopfield. Biomed Res Int 2014:580491. https://doi.org/10.1155/2014/580491. Epub 2014 Jun 16. PMID: 25028662; PMCID: PMC4084588
5. Rinaldi A (2011) When life gets physical. EMBO Rep 13(1):24–7. https://doi.org/10.1038/embor.2011.236. PMID: 22173031; PMCID: PMC3246260
6. Babcock NS et al (2024) Ultraviolet superradiance from mega-networks of tryptophan in biological architectures. J Phys Chem B. https://doi.org/10.1021/acs.jpcb.3c07936
7. Aiello C (2023) It's time to take quantum biology research seriously. Phys Mag. https://www.googletagmanager.com/ns.html?id=GTM-PNZNVC2P
8. Streltsov K, Adesso G, Plenio MB (2017) Quantum coherence as a resource. Rev Mod Phys 89(4):041003

Part II
Reinterpretations

Chapter 5
Uncertainty Principle and Loss of Information

Abstract This chapter reinterprets the uncertainty principle through the lens of the Unified Quantum Object (UQO) framework, proposing that uncertainty is not intrinsic but emergent from deeper, deterministic antecedent structures. These structures, residing in meta-layers ($M1$, $M2$, $M3$), encode deterministic relationships between variables, which project onto the observable B-Layer as uncertain observables. This projection involves information loss, analogous to dimensionality reduction or data compression, giving rise to probabilistic behavior. The degree of uncertainty is modulated by the influence of these meta-layers, potentially reduced by factors like entanglement. Therefore, the uncertainty principle is re-conceptualized as a consequence of this information loss during projection, suggesting that quantum computing architectures designed to access and leverage these antecedent structures could minimize information loss, improve measurement accuracy, and potentially enable a higher degree of deterministic quantum computation.

In this chapter, the uncertainty principle is reinterpreted [1][1] based on the Unified Quantum Object (UQO) construct introduced in Chap. 4. It is the antecedent structure in the UQO that underlies and gives rise to uncertainty. Hence, uncertainty is reframed—not as an intrinsic, irreducible property of quantum systems but as a phenomenon emergent from deeper, more fundamental dynamics.

In its standard form, the uncertainty principle reflects the limits of simultaneous knowledge about conjugate variables, such as position and momentum. This is summarized by Eq. (5.1), *Uncertainty Principle*:

$$\Delta p \Delta x \geq \frac{h}{4\pi}. \tag{5.1}$$

At the antecedent level, a deterministic architecture governs the relationships between the variables x and p. These variables, as perceived at the B-Layer, are emergent approximations or projections of this deeper deterministic interaction: that

[1] An IEEE article [1] initially surfaced a reinterpretation of Heisenberg's uncertainty principle. This chapter builds on it to also create a more sophisticated mathematical rendering of the reinterpretation.

P. Malik, *Pioneering New Avenues in Quantum Technology*, Studies in Smart Technologies, https://doi.org/10.1007/978-981-96-5463-5_5

is, seed configurations at $M1$ encode the deterministic potential that later manifests as uncertain observables at B. The probabilistic nature of uncertainty arises not from inherent randomness but from the projection or collapse of deterministic antecedent structures onto the emergent B-Layer. This projection involves the loss of information or the simplification of higher-layer dynamics, much like the compression of multidimensional data into lower dimensions.

At the B-Layer uncertainty is fully expressed due to the emergent nature of observable properties and measurement constraints. At $M1$, the deterministic structure (seed configurations) begins to constrain the uncertainty, reducing its range. At $M2$, pure properties further refine the determinism so that its influence through m-entanglement may reduce the overall uncertainty in a system based on the prevalence of the pure property in seeds across the computational system. At $M3$, if the complex agglomerate is the primary vehicle of computation, than likely the overall system uncertainty would further decrease. This general idea is summarized by Eq. (5.2), *Generalized Uncertainty Across Layers*:

$$\Delta Q_1 \Delta Q_2 \geq U_L. \tag{5.2}$$

Here Q_1, Q_2 are layer-specific variables (e.g., position and momentum at B-Layer, seed properties at $M1$, pure properties at $M2$, holistic properties at $M3$), U is uncertainty, and $U_B > U_{M1} > U_{M2} > U_{M3}$.

The traditional uncertainty relation might be extended to incorporate the multi-layered structure of the QDC as depicted by Eq. (5.3), *Uncertainty with QDC Adjustment*:

$$\Delta Q_1 \Delta Q_2 \geq \frac{h}{4\pi} f(M1, M2, M3). \tag{5.3}$$

Here $f(M1, M2, M3)$ is a function that accounts for the influence of meta-layers on the base layer uncertainty. This function could potentially reduce the lower bound of uncertainty depending on the deterministic information provided by higher layers. The important point is that something of the meta-layer has to be perceived in Q_1 and Q_2 at the B-layer. If we assume these to be properties in a seed configuration, then (5.3) is suggesting that while there is a bound to the limit of accuracy in distinguishing two conjugate properties that are part of a single seed configuration, this bound may be reduced by the kind of quantum computing architecture employed that would provide insight into the antecedent deterministic structure modeled to be behind the B-layer.

Further, the different types of entanglement introduced in the QDC could modify the uncertainty relations in the following way, as depicted by Eq. (5.4), *Modified Uncertainty due to Multi-Layered Entanglement*:

$$\Delta Q_1 \Delta Q_2 \geq \frac{h}{4\pi} \left(1 - E_p - E_o - E_n - E_m\right). \tag{5.4}$$

Here, E_p, E_o, E_n, and E_m represent the strength of *p*-, *o*-, *n*-, and *m*-entanglement respectively, potentially reducing the overall uncertainty at *B*-Layer.

Equations (5.3) and (5.4) can be combined to create Eq. (5.5), *Unified Uncertainty Relation*:

$$\Delta Q_1 \Delta Q_2 \geq \frac{h}{4\pi} - \sum_k (f_k(M_k) + \alpha_k E_k). \quad (5.5)$$

Here $f_k(M_k)$ reflects the meta-layer contributions and E_k represents entanglement strength.

5.1 Loss of Information

Given the Qualified Deterministic Construct summarized by (4.15), and the reduction in loss of information suggested by (5.5), it is clear that unless a quantum computing architecture accounts for the deterministic structure and the kinds of entanglement due to meta-layers impressing upon *B*-Layer quantum dynamics, there is going to be a loss of information [2, 3]. As per the treatise being presented in this book, that loss of information is indicated by the uncertainty principle.

There are several complementary ways that can be leveraged to begin to understand this loss in information from different perspectives such as dimensionality reduction, entropy changes, information compression, and projection operators.

The relationship between antecedent structures ($M1, M2, M3$) and the *B*-Layer can be likened to compressing higher-dimensional data into lower dimensions [4]. Mathematically, this can be expressed by Eq. (5.6), *Information Loss as Dimensionality Reduction*:

$$\mathcal{P} : \mathcal{H}_M \rightarrow \mathcal{H}_B. \quad (5.6)$$

Here, $\mathcal{H}_M$ is the Hilbert space of the meta-layers ($M1$, $M2$, $M3$); $\mathcal{H}_B$ is the Hilbert space of the *B*-Layer; and $\mathcal{P}$ is the projection operator that maps meta-layer information onto the *B*-Layer [5]. Due to this some deterministic properties encoded in $\mathcal{H}_M$ are compressed, leaving behind probabilistic uncertainty in $\mathcal{H}_B$. The loss of information can be represented by Eq. (5.7), *Quantification of Information Loss*:

$$\dim_{\text{Loss}} = \dim(\mathcal{H}_M) - \dim(\mathcal{H}_B). \quad (5.7)$$

Information loss can also be modeled using entropy, which quantifies the uncertainty or randomness of a system [6]. The projection process increases entropy at the *B*-Layer, and is depicted by Eq. (5.8), *Information Loss due to Increase in Entropy*:

$$\Delta S = S_B - S_M. \quad (5.8)$$

Here, S_M is the entropy of the deterministic structure at meta-layers ($M1, M2, M3$); S_B is entropy at the *B*-Layer, reflecting emergent probabilistic uncertainty.

This increase in entropy corresponds to the simplification of higher-layer dynamics and the emergence of probabilistic behavior. The relationship can be linked to the compression of Shannon information [7], depicted by Eq. (5.9), *Compression of Shannon Information*:

$$\Delta I = I_M - I_B. \tag{5.9}$$

Here I_M and I_B are the information content at the meta-layers and *B*-Layer, respectively.

It is also possible to model the projection process as a lossy compression operation, where deterministic information is "compressed" into an emergent probabilistic framework. A simple model for this is depicted by Eq. (5.10), *Lossy Compression Model*:

$$I_B = I_M - \eta. \tag{5.10}$$

Here I_M is total information in the meta-layer; I_B is the retained information at the B-Layer; and η is information loss, representing discarded or unobservable data.

The loss of information can be formalized through projection operator *P*, as depicted by Eq. (5.11):

$$\psi_B = P\psi_M. \tag{5.11}$$

Here, ψ_M is the deterministic quantum state in the meta-layer space; ψ_B is the resultant lower-dimensional probabilistic *B*-layer space; and *P* is the operator that introduces information loss by simplification from the higher to the lower quantum state.

The fidelity [8] of the projection can be quantified using a measure as depicted by Eq. (5.12), *Fidelity of Projection*:

$$F = |\langle \psi_B | P\psi_M \rangle|^2. \tag{5.12}$$

Low fidelity indicates high information loss.

Information loss during the projection of deterministic antecedent structures onto the *B*-Layer can be mathematically modeled as a weighted combination of dimensionality reduction—quantifying the reduction in degrees of freedom, entropy increase—tracking the increase in randomness, compression noise—representing the discarded information, and projection fidelity—evaluating the effectiveness of layer-to-layer mapping, as depicted by Eq. (5.13), Weighted Information Loss:

$$\text{Loss}_{\text{Total}} = \alpha \text{Dim}_{\text{Loss}} + \beta \Delta S + \gamma \eta + \delta(1 - F). \tag{5.13}$$

Here, α, β, γ, and δ are weights to be applied to the individual information loss components.

In quantum computing, minimizing this information loss through enhanced coherence, optimized architectures, and better error correction mechanisms [9] is key to harnessing the deterministic potential of meta-layer dynamics.

5.2 Some Implications for Quantum Computing

This reinterpretation shifts the quantum paradigm since the uncertainty principle becomes an expression of emergent behavior, not a fundamental limitation. Deterministic antecedent structures, operating at meta-layers ($M1$, $M2$, $M3$) guide the manifestation of quantum phenomena. Such a shift not only enriches our understanding of quantum systems but also paves the way for transformative advances in quantum technologies, rooted in a deeper comprehension of the antecedent structures that underpin reality. While there is an effective decrease in uncertainty due to the deterministic meta-level influence and entanglement, there is also a loss of information in the projection of such determinism onto the B-layer. Hence, Eqs. (5.5) and (5.13) have to be combined to yield Eq. (5.14), *Unified Uncertainty Relation with Loss of Information*:

$$\Delta Q_1 \Delta Q_2 \geq \frac{h}{4\pi} - \left(\sum_k (f_k(M_k) + \alpha_k E_k)\right) + (\alpha \mathrm{Dim}_{\mathrm{Loss}} + \beta \Delta S + \gamma \eta + \delta(1 - F)). \tag{5.14}$$

Reframing uncertainty as emergent has profound implications for the theoretical and practical understanding of quantum systems. Uncertainty in measurement reflects the loss of antecedent information during the transition from deeper structures ($M1$, $M2$, $M3$) to the emergent B-Layer. This suggests that quantum measurements could be improved by accessing or leveraging antecedent structures. Entanglement, typically treated as a fundamental quantum phenomenon, could be seen as a remnant of antecedent deterministic relationships. Zero-point energy [10], often attributed to quantum fluctuations, could instead reflect the "trace" of deterministic dynamics from antecedent layers.

If uncertainty emerges from an antecedent structure, quantum computing architectures can be designed to access and leverage this structure, enabling deterministic algorithms that bypass the probabilistic constraints of conventional quantum systems. Qibits (quaternary quantum bits) proposed in the treatise could encode antecedent structures, allowing for deterministic and layer-specific computations. Many errors in quantum computing arise from the probabilistic nature of quantum systems. Accessing antecedent deterministic dynamics could significantly reduce or eliminate such errors.

5.3 Further Reading

- "Uncertainty: Einstein, Heisenberg, Bohr, and the Struggle for the Soul of Science" by David Lindley: A historical and conceptual exploration of the development of the uncertainty principle.
- "The Uncertainty Principle" by Werner Heisenberg: Heisenberg's own account of the development of the uncertainty principle, offering historical and philosophical insights.
- "Elements of Information Theory" by Thomas M. Cover and Joy A. Thomas: A comprehensive textbook on information theory, covering topics such as entropy, compression, and channel capacity.
- "Information, Physics, Quantum: The Search for Links" by Vlatko Vedral: Explores the deep connections between information theory and physics, including quantum mechanics.
- "Principles of Quantum Mechanics" by P.A.M. Dirac: A classic and rigorous treatment of the mathematical foundations of quantum mechanics.
- "Quantum Theory" by Peter R. Holland: A comprehensive treatment of Bohmian mechanics and its implications.
- "The Quantum Vacuum: An Introduction to Quantum Electrodynamics" by Peter W. Milonni: This book provides a comprehensive overview of the quantum vacuum and zero-point energy.

References

1. Malik P (2021) A light-based Interpretation of Schrodinger's wave equation and Heisenberg's uncertainty principle with implications on quantum computation. In: 2021 IEEE international IOT, electronics and mechatronics conference (IEMTRONICS), Toronto, ON, Canada, pp 1–6. https://doi.org/10.1109/IEMTRONICS52119.2021.9422517
2. Nielsen MA, Chuang IL (2010) Quantum computation and quantum information. Cambridge University Press, Cambridge
3. Horodecki R, Horodecki P, Horodecki M, Horodecki K (2009) Quantum entanglement. Rev Mod Phys 81(2):865–942
4. Lloyd S, Mohseni M, Rebentrost P (2014) Quantum principal component analysis. Nat Phys 10(9):631–633
5. Peres A (2002) Quantum theory: concepts and methods. Kluwer Academic Publishers, New York
6. Wilde MM (2017) Quantum information theory. Cambridge University Press, Cambridge
7. Schumacher B (1995) Quantum coding. Phys Rev A 51(4):2738
8. Jozsa R (1994) Fidelity for mixed quantum states. J Mod Opt 41(12):2315–2323
9. Gottesman D (2010) An introduction to quantum error correction and fault-tolerant quantum computation. Proc Symp Appl Math 68:13–58
10. Chown M (2008) What is the zero-point energy, and does it really exist?. Sci Am. https://www.scientificamerican.com/article/follow-up-what-is-the-zer/

Chapter 6
Quanta and Quantum Object Code

Abstract This chapter introduces the concept of a "Quantum Object Code," a fourfold structure based on quantized space, time, energy, and gravity that governs the emergence and behavior of quantum objects. This code arises from the interplay between the speed of light (c) and Planck's constant (h), proposing an inverse proportionality where h defines a fundamental lower limit corresponding to the upper limit of c at the observable layer of reality (B-layer). This relationship is mathematically supported by combining electromagnetic and photoelectric equations, building upon Planck's quantization of energy and Einstein's concept of light quanta. The traditional constancy of c is challenged by evidence suggesting variations, supported by the Heisenberg Uncertainty Principle, theoretical meta-levels, and research on phenomena like the Scharnhorst effect, as discussed by Feynman and others. These variations in c are linked to wave-particle duality and the stability maintained by h. Crucially, this model posits that space, time, energy, and gravity are emergent and quantized due to the finite nature of c, forming the very basis of the proposed Quantum Object Code and providing a structured framework for understanding quantum phenomena.

Light, traveling at a speed of 186,000 miles per second in a vacuum (c), takes a finite amount of time to traverse any given distance. This implies that a "unit" of light requires a specific amount of time to travel from point A to point B. Quanta at the subatomic level can be thought of as connected to this finite time and distance required for a unit of light to manifest. Planck's discovery—that energy at the subatomic level is quantized and requires a minimum threshold or "quantum" to express itself—aligns intuitively with this concept of a unit of light [1].[1]

It is important to note, however, that Planck originally introduced quanta as a mathematical construct. This approach enabled the derivation of an equation that accurately described the radiation curve of different wavelengths emitted by a heated black body [2]. In contrast, Einstein proposed that quanta represent an intrinsic property of light itself [3], rather than something arising purely from its interaction

[1] This initial idea and the subsequent development of the formulation of space, time, energy, and gravity quantization appeared in [1]. Here the math and narrative have been further developed.

P. Malik, *Pioneering New Avenues in Quantum Technology*, Studies in Smart Technologies, https://doi.org/10.1007/978-981-96-5463-5_6

with matter, as Planck had suggested. Einstein's theory led to the formulation of the photoelectric effect, which established that the energy of emitted electrons depends on the frequency of light. This groundbreaking work earned Einstein the Nobel Prize [4].

If c is perceived as some significant upper limit of the B-layer, then as per this intuition, the lower limit h (Planck's constant) should be inversely proportional to c. This is depicted by Eq. (6.1), *Intuitive Relationship Between h and c*:

$$h \propto \frac{1}{c}. \tag{6.1}$$

This relationship is substantiated by combining two well-known equations: the first is the electromagnetic equation connecting speed of light with wavelength (λ) and frequency (ν), $c = \nu\lambda$, and the second is Einstein's photoelectric equation connecting energy (E) with frequency of light, $E = h\nu$. When these are combined it yields Eq. (6.2), *Verified Relationship Between h and c*:

$$h = \frac{E\lambda}{c}. \tag{6.2}$$

H. A. Lorentz, the Dutch scientist, observed [5] that the constant h serves as the foundation for explaining a wide range of phenomena. It provides the basis for understanding the intensity of radiation and the wavelength at which it reaches its maximum, as well as interpreting numerous quantitative relationships in physics. These include the specific heat of solids, the photo-chemical effects of light, the orbits of electrons within atoms, the wavelengths of spectral lines, the frequency of X-rays produced by electron impacts at specific velocities, the rotational velocities of gas molecules, and the distances between particles in a crystal structure. Lorentz emphasized that quantum conditions play a central role in maintaining the stability of matter and preventing it from dissipating its energy entirely through radiation.

There is emerging evidence suggesting that the speed of light, traditionally considered a constant, may exhibit variations around c, potentially exceeding or falling below this value:

- The Heisenberg Uncertainty Principle implies that photons can travel at varying speeds, occasionally surpassing c for brief durations.
- The concept of alternate space–time realities, referred to as meta-levels in this treatise, suggests that light may travel differently within layers beyond the four-dimensional space–time framework that defines our observable universe [6].
- In *QED: The Strange Theory of Light and Matter* [7] Feynman explains that "there is also an amplitude for light to go faster (or slower) than the conventional speed of light." He elaborates that photons do not always travel in straight lines or strictly at c. Supporting this, research by Scharnhorst at Humboldt University [8], based on quantum electrodynamics, has proposed the possibility of "faster-than-light" photons, a phenomenon referred to as the Scharnhorst effect.

- Perkowitz, in his book *Slow Light* [9], points out that the theory of relativity does not prohibit particles from moving at or beyond the speed of light.

This proposed variance in the speed of light offers deeper insights into the wave-particle duality observed in the quantum realm:

- At the *B*-layer, the speed of light in a vacuum, c, is finite at 186,000 miles per second. If c is assumed to represent a fundamental upper limit at this layer, its inverse, $1/c$, must define a fundamental lower limit.
- This connection is evident in Planck's constant, h, which is proportional to $1/c$, as previously discussed. Planck's constant plays a critical role in sustaining matter by fundamentally limiting the dispersion of energy, as Lorentz has suggested.
- In this context, a wave can be understood as the dematerialization of a particle. For this to occur, h must effectively become a fraction of itself, meaning that the fundamental stability of matter is disrupted, causing it to disperse as a wave. Simultaneously, this would require the speed of light to exceed c, given the inverse proportionality between h and c.
- This relationship further suggests that as information, expressed as a wave, gathers and organizes itself to manifest as matter, it reaches a critical threshold where it condenses into discrete quanta.

Quanta, then can be mathematically expressed in the following manner. Note that space, time, energy, gravity that precipitate in the Unified Quantum Object (UQO) at $|\psi_c\rangle$ as summarized by (4.5), is a macro-level phenomenon and are a symmetrical face of knowledge, power, presence, and harmony, respectively. The emergence of quanta is now seen at the micro-level where seeds that show up in space can belong to any combination of elements in the sets that exist at $|\psi_k\rangle$.

6.1 The Quantization and Structure of Space

The quantization of space, hence, is modeled in the following way as specified by Eq. (6.3), Space Quantization, which derives from (4.8):

$$\text{Space}_{\text{quantization}} = h\left(aX + \sum_{i=0}^{n} b_i Y_i\right). \tag{6.3}$$

Here, $X, Y \in [S_{\text{Pr}}, S_{\text{Po}}, S_K, S_H]$; a and b are integers such that $a > b$; and extending Lorentz's observation, h—Planck's constant, is assumed to be universal.

In this model, 'space' as an emergent property is structured by infinite seeds that can derive from any set. But because the emergence is taking place in a layer of reality generally itself structured by a finite speed of light, c, implying that quantization is a fundamental aspect of its reality, it has to be quantized. The quantization assures that the subtle information is not dissipated, but can accumulate to create seeds, and therefore, the structure of space itself. Such a structure of space appears consistent with recent models on the structure of space as explored in Rovelli's 'Reality Is Not What It Seems' [10].

The structure of space itself would be the summation of all the seeds as in Eq. (6.4), Structure of Space:

$$\text{Space}_{\text{structure}} = \sum_{j=1}^{m} h\left(a_j X_j + \sum_{i=0}^{n} b_{i,j} Y_{i,j}\right). \tag{6.4}$$

Here, m represents the total number of seeds that collectively form space; $X_j \in [S_{\text{Pr}}, S_{\text{Po}}, S_K, S_H]$ is the j-th seed's primary quantization factor; $Y_{i,j} \in [S_{\text{Pr}}, S_{\text{Po}}, S_K, S_H]$ represents the i-th secondary quantization factor associated with the j-th seed; and a_j and $b_{i,j}$ are integers, with the condition $a_j > b_{i,j}$.

6.2 The Quantization of Time

The quantization of time, which holds the inevitability of the seeds that structure space emerging in a phased maturity, hence, is modeled in the following way as specified by Eq. (6.5), Time Quantization:

$$\text{Time}_{\text{quantization}} = h\left(aX + \sum_{i=0}^{n} b_i Y_i\right).\Theta\left(\text{Space}_{\text{quantization}}\right). \tag{6.5}$$

Here, $X \in [S_{\text{Pr}}, S_{\text{Po}}, S_K, S_H]$; $Y \in [S_{\text{Pr}}, S_{\text{Po}}, S_K, S_H]$; a and b are integers such that $a > b$; and extending Lorentz's observation, h—Planck's constant, is assumed to be universal; and $\text{Space}_{\text{quantization}}$ and $\text{Time}_{\text{quantization}}$ take place together as specified by $\Theta\left(\text{Space}_{\text{quantization}}\right)$.

In this model, 'time' as an emergent property of light is structured by an inevitable process of maturity. But because the emergence is taking place in a layer of reality itself structured by a finite speed of light, c, it has to be quantized. The quantization assures that the power is not dissipated but can accumulate to express phased maturity, and therefore, the structure of time itself.

6.3 The Quantization of Energy

Energy, which through a process of containment can result in matter, hence, is modeled in the following way as specified by Eq. (6.6), Energy Quantization:

$$\text{Energy}_{\text{quantization}} = h\left(aX + \sum_{i=0}^{n} b_i Y_i\right).\Theta\left(\text{Space}_{\text{quantization}}\right). \tag{6.6}$$

Here, $X \in [S_{\text{Pr}}, S_{\text{Po}}, S_K, S_H]$; $[Y \in S_{\text{Pr}}, S_{\text{Po}}, S_K, S_H]$; a and b are integers such that $a > b$; and extending Lorentz's observation, h—Planck's constant, is assumed to be universal; and $\text{Space}_{\text{quantization}}$ and $\text{Time}_{\text{quantization}}$ and $\text{Energy}_{\text{quantization}}$ take place together as specified by $\Theta(\text{Space}_{\text{quantization}})$.

In this model, 'energy' as an emergent property results in the reality of matter. But because the emergence is taking place in a layer of reality generally itself structured by a finite speed of light, c, it has to be quantized. The quantization assures that the energy is not dissipated but can accumulate to create matter.

6.4 The Quantization of Gravity

Gravity, which holds seemingly distinct objects in the layer of reality created by c together in a harmony, hence, is modeled in the following way as specified by Eq. (6.7):

$$\text{Gravity}_{\text{quantization}} = h\left(aX + \sum_{i=0}^{n} b_i Y_i\right).\Theta(\text{Space}_{\text{quantization}}). \tag{6.7}$$

Here, $X \in [S_{\text{Pr}}, S_{\text{Po}}, S_K, S_H]$; $Y \in [S_{\text{Pr}}, S_{\text{Po}}, S_K, S_H]$; a and b are integers such that $a > b$; and extending Lorentz's observation, h—Planck's constant, is assumed to be universal; and $\text{Space}_{\text{quantization}}$ and $\text{Time}_{\text{quantization}}$ and $\text{Energy}_{\text{quantization}}$ and $\text{Gravity}_{\text{quantization}}$ take place together as specified by $\Theta(\text{Space}_{\text{quantization}})$.

In this model, 'gravity' as an emergent property results in a harmonious collectivity of seemingly independent objects. But because the emergence is taking place in a layer of reality itself structured by a finite speed of light, c, it has to be quantized. The quantization assures that the harmony is not dissipated but can accumulate to express more and more complex collectivities on large-scale.

6.5 Quantum Object Code

The preceding analysis also suggests that space is a repository of pre-genetic information or of pre-genetic libraries. Time, gravity, and energy work with space to bring about material possibilities. Change to any material possibility requires therefore the action of space, time, energy, and gravity. At the margin, there is a fourfold quantization that brings about such change.

Space, time, energy, and gravity in this perspective are not absolute but are emergent, as initially suggested by (2.4) that summarized the Hilbert space where light travels at speed c. It was also proposed that any quantum object that arises in this Hilbert space will have fourfold space–time–energy–gravity quantum object code that will specify how a quantum object shows up and behaves based on that code. The code is based on Eqs. (6.3), (6.5), (6.6), and (6.7) that correspond to S_q, T_q, E_q, G_q, respectively, and an equation that combines these, Eq. (6.8), Quantum Object Code is as follows:

$$Q_{\text{object}} = \left(S_q \Lambda T_q \Lambda E_q \Lambda G_q\right). \tag{6.8}$$

The full Quantum Object Code Q_{object} encapsulates how a quantum object emerges and behaves, governed by the interplay of quantized space, time, energy, and gravity, and will be further taken up in Chap. 10 where simulation of a simple hypothetical atom will be considered. Each component interacts through h, with X and Y_i reflecting the object's specific dimensional context. This unified perspective highlights how quantum objects are dynamic entities influenced by multidimensional factors, providing a structured yet flexible framework for understanding quantum phenomena.

6.6 Further Reading

- "'Subtle Is the Lord…': The Science and the Life of Albert Einstein" by Abraham Pais. A highly regarded biography of Einstein.
- Black-Body Theory and the Quantum Discontinuity, 1894–1912. Thomas S. Kuhn: A detailed historical analysis of the development of quantum theory.
- "Faster Than Light: Superluminal Loopholes in Physics" by Nick Herbert: This book explores various theoretical possibilities for exceeding the speed of light.
- "Quantum Field Theory in a Nutshell" by A. Zee: A comprehensive and relatively accessible introduction to quantum field theory, which includes QED as a specific case.
- "Spacetime Physics" by Edwin F. Taylor and John Archibald Wheeler: A classic textbook on special relativity with a focus on conceptual understanding.
- "The Fabric of the Cosmos: Space, Time, and the Texture of Reality" by Brian Greene: Explores the nature of space, time, and the universe, including discussions of string theory and extra dimensions.

References

1. Malik P. 20-book cosmology of light series. Amazon Author Page. Accessed 14 Jan 2025. https://www.amazon.com/author/pravirmalik
2. Planck M (1900) Zur Theorie des Gesetzes der Energieverteilung im Normalspektrum [On the theory of the law of energy distribution in the normal spectrum]. Verh Dtsch Phys Ges 2:237–245

3. Einstein A (1905) Über einen die Erzeugung und Verwandlung des Lichtes betreffenden heuristischen Gesichtspunkt [On a heuristic viewpoint concerning the production and transformation of light]. Ann Phys 322(6):132–148
4. Nobel Prize Outreach AB (1921) The Nobel prize in physics. Accessed 11 Jan 2025. https://www.nobelprize.org/prizes/physics/1921/summary/
5. Glazebrook RT (1928) Hendrik Antoon Lorentz. Nature 121(3043):287–289
6. Hawking SW (1988) A brief history of time: from the big bang to black holes. Bantam Books, New York
7. Feynman RP (1985) QED: the strange theory of light and matter. Princeton University Press, Princeton, NJ
8. Chown M (1990) Can photons travel faster than light? New Sci 126(1711):32
9. Perkowitz S (2004) Slow light: science and applications. Elsevier Academic Press, Amsterdam
10. Rovelli C (2017) Reality is not what it seems: the journey to quantum gravity. Translated by Simon Carnell and Erica Segre. Riverhead Books, New York

Chapter 7
Schrodinger's Equation and Evolution of Quantum Systems

Abstract This chapter reinterprets the Schrödinger equation within the multi-layered UQO quantum model, shifting from its traditional role as a description of wavefunction evolution to a mechanism for translating information from antecedent layers into material existence. While the standard equation is fundamental to quantum mechanics, describing probabilistic quantum behavior and playing a key role in quantum computation, its origins and implications have been subject to various interpretations, from hidden variable theories to the many-worlds interpretation. This work proposes a modified Schrödinger equation grounded in the concept of qualified determinism (QD), incorporating the influence of meta-layers (M1, M2, M3) and their associated deterministic, pure property, and complex agglomerate structures. This modified "Unified Dynamics Schrödinger Equation" also integrates the emergent properties of space, time, energy, and gravity, framing quantum computation as a creative process aimed at accessing novel functionalities theorized to exist in seed-form within these antecedent layers.

The Schrödinger equation is fundamental to quantum mechanics as it describes how quantum systems evolve over time [1], serving as a cornerstone for understanding and predicting their behavior, much like Newton's laws do for classical systems. It also plays a crucial role in quantum computation, helping to describe qubit behavior, design algorithms, simulate quantum systems, analyze hardware, and develop error correction. Unlike the deterministic framework of classical mechanics, the Schrödinger equation highlights the probabilistic nature of the quantum world, enabling the calculation of the likelihood of various outcomes rather than definitive certainties. In this chapter, the equation will be modified and reinterpreted [2][1] based on the model presented in this book.

[1] An IEEE article [2] initially surfaced a reinterpretation of Schrodinger's wave equation. This chapter builds on it to also create a more sophisticated mathematical rendering of the reinterpretation.

P. Malik, *Pioneering New Avenues in Quantum Technology*, Studies in Smart Technologies, https://doi.org/10.1007/978-981-96-5463-5_7

7.1 Reinterpretation of Schrodinger's Equation

Here, the time-dependent form of Schrodinger's equation is used in preference to its time-independent form because it can be applied to a broader range of problems, including those where the potential energy or other conditions are changing over time. As covered in the previous chapters, the quantum model being developed in this treatise has many conditions that change over time due to the dynamics of meta-layers and existence of quantum object code that creates a different degree of dynamism. Restating the simplified time-dependent form of Schrödinger's Eq. (7.1):

$$i\frac{h}{4\pi}\frac{\partial}{\partial x}\psi = \widehat{H}\psi. \tag{7.1}$$

Here, ψ represents a wave form and can be understood as a probabilistic cloud of possible states. $\widehat{H}$ is the Hamiltonian operator, which serves as a focusing function. Essentially, this equation may suggest that the way a wave form evolves over time corresponds to an expressible state of the inherent possibilities within the cloud of potential states. When analyzed in the context of the multi-layered model the "cloud of possible states" can be seen as another way to articulate how 'form' exists beyond its physical manifestation.

The equation alludes to the profound mystery involved in unveiling the material realm. Various interpretations of the quantum world attempt to explore the potentiality implicit within this equation. As summarized by Kleinman in The Four Faces of the Universe [3], these interpretations range from theories addressing "hidden variables," which highlight our ignorance of the system's deeper aspects, to David Bohm's concept of the "implicate order," wherein every element contains within it the totality of the universe [4]. Another notable perspective is the "many-worlds" interpretation, which posits that each quantum observation leads to the splitting of the universe into parallel and disconnected realities [5]. Richard Feynman, in his lectures on Schrödinger's equation, is reported to have said something to the effect that it was not possible to derive it from anything that one may know and that the equation came out of the mind of Schrodinger [6]. The Copenhagen Interpretation of Quantum Mechanics [7] notably bypasses inquiries into the nature of reality itself, instead focusing solely on the observed universe. This orientation is reflective of Niels Bohr's stance [8], one of the interpretation's founders, who famously asserted that addressing the nature of reality was not the purpose of physics.

Considering Schrödinger's equation, it is significant to note that the constant 'i' represents a complex number, suggesting an interplay between two dimensions: one "real" and the other "imaginary." The "imaginary" dimension can be interpreted as the meta-levels intrinsic to the mathematical model discussed in this treatise. Furthermore, $\frac{h}{4\pi}$ aligns with the suggestion that h (Planck's constant) may fractionally diminish as c (the speed of light) increases as explored in the previous chapter. The wavefunction's rate of change, $\frac{\partial}{\partial x}\psi$, is inherently linked to both i and $\frac{h}{4\pi}$. Its full significance emerges within the context of $ix\frac{h}{4\pi}x\frac{\partial}{\partial x}\psi$, a dynamic essential for integrating the behavior of multiple layers of the model. Additionally, $\frac{\partial}{\partial x}\psi$ is intimately related to $\hat{H}\psi$, which implies the presence of systemic "energy" represented by the Hamiltonian operator, $\hat{H}$. When applied to the wavefunction ψ, the Hamiltonian forecasts the manner in which the wave will evolve.

In this perspective, "form" can be viewed as existing in its essence as the outcome of dynamics at the meta-layers. Schrödinger's equation thus becomes a mechanism for arbitrating information from beyond the quantum veil—residing in antecedent layers—and translating it into material existence. Consequently, quantum computation emerges as a predominantly creative endeavor, as opposed to a solely constructive process. Unlike classical computational methods that rely on predefined programming instructions, quantum computation can aim to create something entirely new. This newness corresponds to enhanced or unique functions theorized to exist within the antecedent layers of the quantum model.

7.2 A Unified Dynamics Schrodinger Equation

Using the treatise's perspective on qualified determinism (QD), Schrödinger's Wave Equation can be expanded to incorporate the layered structure and deterministic features described. The reinterpretation integrates antecedent structures and the contribution of the emergent space, time, energy, gravity properties in Eq. (7.2), *Unified Dynamics Schrodinger Equation*:

$$i\frac{h}{4\pi}\frac{\partial}{\partial x}\psi_B = \left(H_B + \sum_{k=S,T,E,G} h\left(x_k a_k + \sum_{i=0}^{n} Y_{i,k} b_{i,k}\right)\right)\psi_B + \sum_{i-1}^{3} f_i\left(H_{M_i}\psi_{M_i}\right). \tag{7.2}$$

Here:

- $H_B\psi_B$ governs the wavefunction's intrinsic behavior at the base layer.
- $\sum_{k=S,T,E,G} h\left(x_k a_k + \sum_{i=0}^{n} Y_{i,k} b_{i,k}\right)\psi_B$ accounts for how space, time, energy, gravity quantization modify ψ_B.
- $\sum_{i-1}^{3} f_i\left(H_{M_i}\psi_{M_i}\right)$ represents the contribution of meta-layers (M1, M2, M3) to the evolution of the wavefunction.

The equation can be further modified by incorporating the effects of entanglement as depicted by Eq. (7.3), *Unified Dynamics Schrodinger Equation with Entanglement*:

$$i\frac{h}{4\pi}\frac{\partial}{\partial x}\psi_B = \left(H_B + \sum_{k=S,T,E,G} h\left(x_k a_k + \sum_{i=0}^{n} Y_{i,k} b_{i,k}\right)\right)\psi_B$$

$$+\sum_{i-1}^{3} f_i\left(H_{M_i}\psi_{M_i}\right) - \sum_{j=p,o,n,m} \alpha_j E_j \psi_B. \tag{7.3}$$

Here:

- $\sum_{j=p,o,n,m} \alpha_j E_j \psi_B$ models entanglement-related factors influencing ψ_B.

  Note that (7.2) is the general form that will be used in the rest of the book.

7.3 Further Reading

- "Atomic Physics and the Description of Nature" by Niels Bohr: This is a key essay where Bohr discusses his views on the nature of physics and the limitations of classical concepts in the quantum realm.
- "The Philosophy of Niels Bohr" by Aage Petersen: This article provides insights into Bohr's philosophical views. It includes the often-quoted statement: "It is wrong to think that the task of physics is to find out how nature is. Physics concerns what we can say about nature."
- "Physics and Philosophy: The Revolution in Modern Science" by Werner Heisenberg: Heisenberg, a close collaborator of Bohr, discusses the philosophical implications of quantum mechanics and the Copenhagen interpretation.
- "The Description of Nature: Niels Bohr and the Philosophy of Quantum Physics" by John Honner: A scholarly analysis of Bohr's philosophical approach to physics.
- "Quantum Theory: Concepts and Methods" by Asher Peres: A comprehensive textbook that covers various interpretations of quantum mechanics, including the Copenhagen interpretation, many-worlds interpretation, and Bohmian mechanics.
- "Many Worlds? Everett, Quantum Theory, & Reality" edited by Simon Saunders, Jonathan Barrett, Adrian Kent, and David Wallace: A collection of essays exploring the many-worlds interpretation in detail.
- "Emergence: From Chaos to Order" by John H. Holland: A classic book on the concept of emergence and its role in complex systems.
- "More Than the Parts: Holism and Nonseparability in Science, Philosophy, and Society" edited by Michael Silberstein and Jeff McGeever: Explores the concept of holism and its implications for various fields, including physics.

References

1. Britannica (2025) Schrödinger equation. https://www.britannica.com/science/Schrodinger-equation. Accessed Accessed 11 Jan 2025
2. Malik P (2021) A light-based interpretation of Schrodinger's wave equation and Heisenberg's uncertainty principle with implications on quantum computation. In: 2021 IEEE International

IOT, Electronics and Mechatronics Conference (IEMTRONICS), Toronto, ON, Canada, pp 1–6. https://doi.org/10.1109/IEMTRONICS52119.2021.9422517
3. Kleinman G (2006) The four faces of the universe: an integrated view of the cosmos. Random House, New York
4. Bohm D (1983) Wholeness and the implicate order. Routledge, London
5. DeWitt BS, Graham N (eds) (1973) The many-worlds interpretation of quantum mechanics. Princeton University, Princeton
6. Feynman RP (2013) Lectures on quantum mechanics. Edited by Hans Gottlieb. Basic Books, New York
7. Faye J (2024) Copenhagen interpretation of quantum mechanics. In: Stanford Encyclopedia of Philosophy. Stanford University. https://plato.stanford.edu/entries/qm-copenhagen/. Accessed 31 May 2024
8. Bohr N (1949) Discussion with Einstein on epistemological problems in atomic physics. In: Schilpp PA (ed) Albert Einstein: philosopher-scientist. Library of Living Philosophers, Evanston, IL, pp 201–241

Chapter 8
Reimagining Euler's Endless Wave

Abstract This chapter reinterprets Euler's Identity as a dynamic framework for understanding quantum processes, moving beyond its conventional mathematical significance to explore its implications for quantum computation. Traditionally celebrated for uniting fundamental mathematical constants (0, 1, *i*, *e*, and π), Euler's Identity is here viewed as representing the transition from abstract potentiality to material manifestation within a multi-layered conceptual space modeled by the Universal Quantum Object (UQO). The constants *e* and π are associated with key transitional dynamics: *e* with exponential experimentation between differentiation and stability, and π with the dynamic interplay between stability and manifestation. The imaginary unit *i* acts as a bridge between conceptual and real domains. In this context, the undulating wave described by Euler's formula is not merely a probabilistic amplitude but a mechanism mediating between layers of possibility. This reinterpretation suggests that quantum computation, mediated by *e*, π, and *i*, is not solely a probabilistic framework but a creative process that recursively transforms abstract potential into tangible reality, with the qibit acting as the mediator of this constant negotiation. This perspective offers a novel understanding of quantum computation as a process of dynamic arbitration between conceptual layers, challenging conventional views of the field.

Recognized for its elegance, Euler's identity, Eq. (8.1), is widely celebrated as one of the most aesthetically pleasing results in mathematics [1]. Its allure arises from the way it joins five fundamental constants—0, 1, *i*, *e*, and π—into a concise symbolic statement. Here, *i* refers to the imaginary unit, $e \approx 2.71828$ is the foundation of natural logarithms, and $\pi \approx 3.14159$ is defined by the ratio of a circle's circumference to its diameter. The concise integration of these elements underscores the concept of mathematical beauty:

$$e^{i\pi} + 1 = 0. \tag{8.1}$$

Equation (8.1) also illustrates deeper processes of possibility, differentiation, stability, and manifestation within quantum systems. The "wave" aspect of Euler's

P. Malik, *Pioneering New Avenues in Quantum Technology*, Studies in Smart Technologies, https://doi.org/10.1007/978-981-96-5463-5_8

formula [2] is shown in Eq. (8.2), *Wave-Aspect of Euler's Formula*:

$$e^{ix} = \cos x + i \sin x \tag{8.2}$$

This sinusoidal form can be seen as more than just an expression for probability amplitudes. It serves as a representation of a dynamic mechanism that unites different layers of possibility within conceptual spaces. These layers—encapsulated by the Universal Quantum Object (UQO) model—trace the transition from abstract potential toward realized phenomena. Within quantum computation, the qibit (see Chap. 18) functions as an intermediary that negotiates among these layers, shaped by the oscillatory behavior. The UQO [reprinted from Eq. (4.5)] appears again in Eq. (8.3):

$$\text{UQO} = \begin{bmatrix} |\psi_\infty\rangle = m(\text{Pr},\ \text{Po},\ \text{K},\ \text{H}) \\ (\downarrow\ \text{P}_k|\psi_\infty\rangle) \\ |\psi_k\rangle = n(S_{\text{Pr}}, S_{\text{Po}}, S_{\text{K}}, S_{\text{H}}) \\ (\downarrow\ \text{P}_g|\psi_k\rangle) \\ |\psi_g\rangle = o(S_{\text{Pr}} \times S_{\text{Po}} \times S_{\text{K}} \times S_{\text{H}}) \\ (\downarrow\ \text{P}_c|\psi_g\rangle) \\ |\psi_c\rangle = p(\text{S}, \text{T}, \text{E}, \text{G}) \end{bmatrix}. \tag{8.3}$$

In the UQO framework, the progression from $|\psi_\infty\rangle$ (all possibilities) to $|\psi_c\rangle$ (material manifestation involving space, time, energy, and gravity) involves stages of quantization, with each stage characterized by the significance of whole numbers:

- 0: All possibility ($|\psi_\infty\rangle$).
- 1: Unity within possibility.
- 2: Differentiation ($|\psi_k\rangle$).
- 3: Stability ($|\psi_g\rangle$).
- 4: Manifestation ($|\psi_c\rangle$).

The constants e and π, central to Euler's identity, highlight the transitional dynamics between these stages [3][1]:

- e (~2.718): Represents exponential growth/decay [4], suggesting how the move from 2 (differentiation) to 3 (stability) involves a continuous, self-referential dynamic. The defining derivative property of the exponential function, shown in Eq. (8.4), captures how this process evolves without losing its underlying form:

$$\frac{\mathrm{d}e^y}{\mathrm{d}x} = e^y. \tag{8.4}$$

[1] An IEEE article [3] suggested how to view Euler's identity in the context of quantum computation. In this chapter, the math has been reformulated to be consistent with the underlying UQO-based quantum mathematics being developed.

- π (~3.14159): Encodes periodic behavior and thus highlights how stability evolves to a state of manifestation, linking 3 (stability) to 4 (full expression) [5]. It reflects the cyclical, oscillatory aspect of quantum systems, which transform a stable form into a more complete material reality through ongoing cycles.

Together, e and π serve as mathematical tools that describe the interplay of growth, periodicity, and transformation across conceptual and material layers in quantum systems. The imaginary unit i, as the bridge between real and conceptual realms, ensures that these transitions remain dynamically interconnected, embodying the creative potential of quantum computation.

In this sense, Euler's identity becomes a conceptual template showing how wave-like oscillations orchestrate the transition from higher-level possibilities into tangible forms. From the perspective of quantum computation, this is not just a probabilistic feature but a fundamentally creative operation—guided by e, π, and i—that persistently transforms abstract possibility into realized phenomena. Euler's identity thus illuminates how differentiation, growth, stability, and manifestation interlock in the quantum realm.

8.1 Significance of Numbers i, 0, 1, 2, 3, and 4

The imaginary unit i (the square root of -1) exemplifies a puzzle at the core of mathematical reality [6]. Although conventional wisdom might label the square root of -1 as "impossible," the existence of i underpins crucial solutions and technological advancements. This paradox—something "unreal" yet vital—sets a precedent for considering the layered approach in this book, where i plays a meaningful but not directly observable role. Hence, Euler's identity triggers a deeper inquiry: what does it represent and how does it reshape our understanding of quantum computation?

Before interpreting Euler's identity, we should first understand the significance of 0, 1, e, and π within it. To do so, we must clarify the "effective range" in which 0, 1, e, and π lie. The UQO model (Eq. 8.3) will guide us through values 0–4, suggesting specific meanings for 0, 1, 2, 3, and 4.

8.1.1 Significance of '0'

In this framework, '0' is mirrored by the condition $|\psi_\infty\rangle$, interpreted as a comprehensive informational state, or "all possibility." This idea is captured by Eq. (8.5), *'All Possibility' as the Significance of '0'*:

$$(\text{‘0’} \equiv |\psi_\infty\rangle = m(\text{Pr},\ \text{Po},\ \text{K},\ \text{H})) \asymp \text{All Possibility}. \tag{8.5}$$

What (8.5) is suggesting is that '0' is identical to ($\equiv$) $|\psi_\infty\rangle$, and further, that the specific interpretation of '0' as suggested by the segment '('0' $\equiv |\psi_\infty\rangle = m(\text{Pr, Po, K, H})$)' is equivalent to ($\asymp$) 'All Possibility'.

8.1.2 Significance of '1'

Further, in $|\psi_\infty\rangle$, there is a oneness or unity that is evident, which becomes the significance of '1'. But as can be seen in Eq. (8.3), 'unity' as the significance of '1', the '1' as 'unity', is a subset ($\subseteq$) of $|\psi_\infty\rangle$ or 'all possibility' and focuses on that aspect only. There are other significances of $|\psi_\infty\rangle$ that are therefore not included in the unity-aspect focused on in Eq. (8.6), *'Unity' as the Significance of '1'*.

Within $|\psi_\infty\rangle$, there is a unifying principle, which suggests the meaning of '1.' As shown in Eq. (8.6), *'Unity' as the Significance of '1'*, unity constitutes a subset of that all-encompassing possibility:

$$(\text{'1'} \subseteq |\psi_\infty\rangle = m(\text{Pr, Po, K, H})) \asymp \text{Unity}. \quad (8.6)$$

Thus, Eq. (8.6) conveys that '1' is a subset of $|\psi_\infty\rangle$ symbolizing a specific perspective of the entire possibility space: unity. In Eq. (8.7), *Relationship between Unity and All Possibility,* a conceptual relationship is established between "unity" and "all possibility":

$$\text{State (Unity)} \subseteq \text{State (All Possibility)} \quad (8.7)$$

which indicates "unity" is contained within "all possibility," yet does not fully exhaust it.

8.1.3 Significance of '2'

When the UQO projection ($\downarrow P_k|\psi_\infty\rangle$) precipitates a state $|\psi_k\rangle$, it reflects a differentiation of implicit properties in $|\psi_\infty\rangle$. This creates four broad sets S_{Pr}, S_{Po}, S_{k}, and S_{H}. This initial separation marks the onset of materialization. Hence, '2' embodies "differentiation," as shown in Eq. 8.8, *Differentiation' as Significance of '2'*:

$$(\text{'2'} \equiv |\psi_k\rangle = \text{n}(S_{\text{Pr}}, S_{\text{Po}}, S_{\text{K}}, S_{\text{H}})) \asymp \text{'Differentiation'}. \quad (8.8)$$

In short, Eq. (8.8) indicates that '2' is identical ($\equiv$) to $|\psi_k\rangle$, and that the associated meaning of '2' aligns ($\asymp$) with "differentiation."

8.1.4 Significance of '3'

Another projection step $(\downarrow \ \mathrm{P}_g|\psi_k\rangle)$ stabilizes differentiation, leading to $|\psi_g\rangle = o(S_{\mathrm{Pr}} \times S_{\mathrm{Po}} \times S_{\mathrm{K}} \times S_{\mathrm{H}})$. By combining attributes from the four sets, $|\psi_g\rangle$ offers a foundational point for eventual materialization, realized when $|\psi_c\rangle = p(\mathrm{S}, \mathrm{T}, \mathrm{E}, \mathrm{G})$, is activated.

Biological evolution [7–9] suggests that such materialization might unfold in steps. This incremental approach underpins the meanings for '3' and '4.' Building on cosmological mathematics [10], fractal research [11], and complex systems innovation [12], one can envision how $|\psi_g\rangle$ produces a seed in three distinct states—conceptual, static, and dynamic—allowing it to be "recorded" in the space–time–energy–gravity interplay of $|\psi_c\rangle$. These three components form a triangle, implying an initial stability where light travels at speed c.

Thus, '3' denotes "stability," as indicated by Eq. (8.9), *'Stability' as Significance of '3'*:

$$\left(\text{'3'} \subseteq \begin{bmatrix} |\psi_g\rangle = o(S_{\mathrm{Pr}} \times S_{\mathrm{Po}} \times S_{\mathrm{K}} \times S_{\mathrm{H}}) \\ (\downarrow \ \mathrm{P}_c|\psi_g\rangle) \\ |\psi_c\rangle = p(\mathrm{S}, \mathrm{T}, \mathrm{E}, \mathrm{G}) \end{bmatrix} \right) \asymp \text{'Stability'}. \tag{8.9}$$

Note further that the 'top' vertex of the envisioned triangle specifying the unique element in essence can be viewed as a play of the property of 'knowledge'. The vertex symbolizing the static aspect can be viewed as a play of the property of 'presence'. The vertex signifying the dynamic aspect can be viewed as a play of the property of 'power'. But the fact that only three properties of light are required for a stable triangle, hence 'stability,' means that '3' can only be a subset ($\subseteq$) of $\left[\left(|\psi_g\rangle = o(S_{\mathrm{Pr}} \times S_{\mathrm{Po}} \times S_{\mathrm{K}} \times S_{\mathrm{H}})\right) \rightarrow \ (|\psi_c\rangle = p(\mathrm{S}, \mathrm{T}, \mathrm{E}, \mathrm{G}))\right]$ simply because the fourth property does not yet appear to exist.

8.1.5 Significance of '4'

'4' extends the notion of '3' by adding the missing property of "harmony." Conceptually, the triangle formed by concept-static-dynamic states is expanded through the presence of an additional element—hence implying 'harmony'. The explicit reflection of all four original properties leads to complete manifestation of possibilities. Therefore, '4' signifies "manifestation," as seen in Eq. (8.10), *'Manifestation' as the Significance of '4'*:

$$\left(\text{'4'} \subseteq \begin{bmatrix} |\psi_g\rangle = o(S_{\mathrm{Pr}} \times S_{\mathrm{Po}} \times S_{\mathrm{K}} \times S_{\mathrm{H}}) \\ (\downarrow \ \mathrm{P}_c|\psi_g\rangle) \\ |\psi_c\rangle = p(\mathrm{S}, \mathrm{T}, \mathrm{E}, \mathrm{G}) \end{bmatrix} \right) \asymp \text{'Manifestation'} \tag{8.10}$$

Equation (8.10) shows that '4' is a larger subset ($\subseteq$) of the same action sequence than '3' and aligns ($\asymp$) with "manifestation." Equation (8.11), *Relationship between Stability and Manifestation*, specifies how stability is nested in manifestation:

$$\text{State(Stability)} \subseteq \text{State (Manifestation)}. \tag{8.11}$$

Thus, stability is contained within manifestation, but manifestation brings additional complexity. There is also a symmetry between '1' and '4': '1' indicates a unification of the four attributes in a conceptual sense, while '4' expresses that totality at the material plane.

8.2 Euler's Identity as a Process of Materialization

From the UQO perspective, we can interpret the significance of 0–4 when discussing Euler's identity (8.1), since e and π numerically appear between 2 and 4. Understanding 2 (differentiation), 3 (stability), and 4 (manifestation) illuminates these constants.

Summarizing, '0' represents the absolute origin of all possibility, and '1' highlights the unity within it. Moving from '1' to '2' introduces differentiation; '3' points to early stable form, and '4' indicates final materialization.

Between '2' and '3' lies e (~2.71828), associated with exponential dynamics. Because the exponential's derivative [13] is itself [Eq. (8.4)], e becomes a metaphor for unending "experimentation" that draws differentiation toward stability. Hence, we may define e in Eq. (8.12), *'Exponential Experimentation' as Significance of 'e'*:

$$\left(\text{`}e\text{'} \equiv \frac{\mathrm{d}\left(e^{\text{`Differentiation'} \rightarrow \text{`Stability'}}\right)}{\mathrm{d}x}\right) \asymp \text{`Exponential Experimentation'}. \tag{8.12}$$

Conceptually, taking the derivative of exponential processes preserves the same exponential character, mirroring how that exploratory dynamic continues even as a stable form emerges.

Likewise, π (~3.14159) occupies a place between '3' (stability) and '4' (manifestation). Thus, π signifies the aspect of cyclical motion through which a stable form can transition to a more realized expression. Equation (8.13), *'Stability Subjected to Dynamism' as Significance of 'π'*, captures this idea:

$$(\text{`}\pi\text{'} \equiv (\text{`Stability'} \rightarrow \text{`Manifestation'}) \asymp \text{`Stability Subject to Dynamism'}. \tag{8.13}$$

In other words, across e, π, and i, Euler's identity (8.1) connects $|\psi_\infty\rangle$ to $|\psi_c\rangle$ through a sequence of quantizations, '0' leads to '1,' to '2,' to '3,' and finally to '4,' all facilitated by i, e, and π.

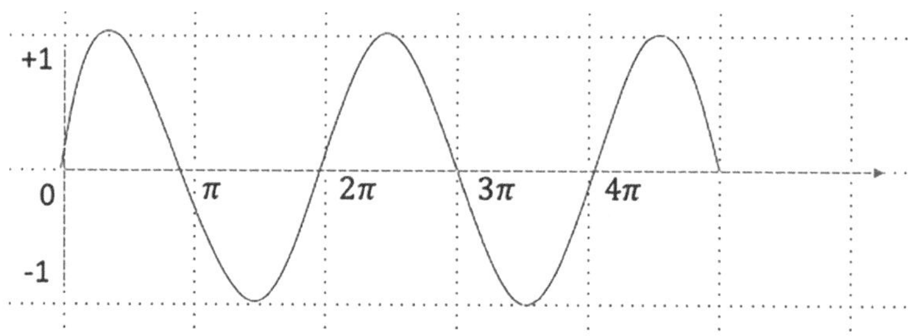

Fig. 8.1 Endless wave

8.3 Some Implications for Quantum Computation

The discussion so far offers a layered interpretation of Euler's identity, linking the constants 0, 1, *i*, *e*, and π with the UQO structure. Depending on one's perspective, Euler's formula [Eq. (8.2)], which undergirds Euler's identity (8.1), can be seen as mapping *e*, *i*, and π (or a variant where π becomes a scaling factor) to a perpetual wave shaped by cosine and sine, connecting real and imaginary domains.

The endless wave is depicted in Fig. 8.1, *Endless Wave.*

Such a "wave" emphasizes that no matter how large the angle grows (horizontal axis), the wave's amplitude (vertical axis) remains bounded. Many quantum theorists consider these bounds—often between 0 and 1—to be ideal for probability interpretations [14]. This perspective underpins the probability-based reading of quantum mechanics, in line with the Copenhagen interpretation, where Euler's formula offers a neat way to code quantum phenomena in a wavefunction.

However, by viewing Euler's identity (8.1) in the UQO context, one sees that the wave is not solely about probability. Instead, it represents a dynamic link among conceptual layers of possibility. In a suitable quantum framework, the qibit is the fundamental node, and its states are shaped by ongoing mediation among these conceptual levels, driven by the wave-like activity.

In essence, this process is a computation. By incorporating integers 0–4 at the material quantum level, we obtain a perspective in which quantum computation is not merely probabilistic; rather, it is a deeply creative phenomenon. It manages the interplay between meta-level conceptual potential and the domain of tangible reality. This is a departure from standard viewpoints and hints that quantum computing might be reimagined altogether, as examined throughout this book.

8.4 Further Reading

- "Dr. Euler's Fabulous Formula: Cures Many Mathematical Ills" by Paul J. Nahin: This book provides a comprehensive and engaging exploration of Euler's formula and its applications in various fields of mathematics and engineering.

- "Imagining Numbers (Particularly the Square Root of Minus Fifteen)" by Barry Mazur: A more philosophical and historical exploration of complex numbers and their significance in mathematics.
- "Helgoland: Making Sense of the Quantum Revolution" by Carlo Rovelli: A more recent book that offers a relational interpretation of quantum mechanics, challenging some aspects of the Copenhagen interpretation.
- "Quantum Mechanics: Historical Contingency and the Copenhagen Hegemony" by James T. Cushing: A historical and critical analysis of the development and influence of the Copenhagen interpretation.
- "A New Kind of Science" by Stephen Wolfram: A controversial but thought-provoking book that explores the idea that simple computational rules can generate complex behavior.
- "Gödel, Escher, Bach: An Eternal Golden Braid" by Douglas Hofstadter: While not directly about physics, this book explores themes of self-reference, recursion, and the relationship between different levels of description, which are relevant to the idea of multi-layered systems.

References

1. Devlin K (2002) The most beautiful equation. Wabash Magazine, Winter/Spring 2002. https://www.wabash.edu/magazine/2002/winterspring2002/mostbeautiful.html
2. Euler L (1988) Introduction to analysis of the infinite. Translated by John D. Blanton. Springer, New York
3. Malik P (2021) A light-based interpretation of Euler's identity with implications on quantum computation. In: 2021 IEEE 12th Annual Information Technology, Electronics and Mobile Communication Conference (IEMCON), Vancouver, BC, pp 0908–0914. https://doi.org/10.1109/IEMCON53756.2021.9623157
4. Weisstein EW (2025) "e." in mathworld—a wolfram web resource. https://mathworld.wolfram.com/e.html. Accessed 12 Jan 2025
5. Euler L (1922) Introduction to analysis of the infinite. In: Collected works, Series 1, vol 8. Teubner, Leipzig
6. Weisstein EW (2025) "i" In MathWorld—a wolfram web resource. https://mathworld.wolfram.com/i.html. Accessed 12 Jan 2025
7. Asano M, Basieva I, Khrennikov A, Ohya M, Tanaka Y (2015) Quantum information biology: from information interpretation of quantum mechanics to applications in molecular biology and cognitive psychology. Found Phys 45(10):1362–1378
8. Vilasini V, Renner R (2021) Causal and compositional structure of unitary transformations. Phys Rev Lett 127(19):190402
9. Zurek WH (2009) Quantum Darwinism. Nat Phys 5(3):181–188
10. Malik P (2024) Cosmology of light illustrated: a mathematical integration of matter, life, history, civilizarion, universe and self. Possibilities Publishing, Civilization
11. Malik P (2015) The fractal organization: creating enterprises of tomorrow. SAGE Publications, New Delhi
12. Malik P (2017) Development and evaluation of a framework for an engine of innovation in complex adaptive systems. PhD thesis, University of Pretoria
13. Maor E (2009) e: the story of a number. Princeton University Press, Princeton
14. Turok N (2012) The universe within: from quantum to cosmos. House of Anansi Press, Toronto

Part III
Quaternary Interpretation of Quantum Dynamics

Chapter 9
Simulating Space, Time, and Probabilities

Abstract This chapter challenges the prevailing probabilistic paradigm in quantum computing, rooted in Feynman's vision of simulating nature through probabilistic quantum computers. Instead, it leverages the Qualified Determinism Construct (QDC), with its intrinsic fourfold structure within quantum space, leading to ordered, albeit infinitely varied, manifestations. While Feynman's approach aims to simulate space, time, and probabilities based on building a probabilistic simulator of a probabilistic nature, the QDC framework reimagines space and time as emergent properties governed by meta-functional logic, replacing a purely probabilistic view with a more structured fourfoldness. This reinforces the need for a new genre of quantum computers based on the interconnected dynamics of a quantum object's fourfold dimensions. By adopting a multi-layered model where emergent phenomena are projections of the four foundational properties of presence, power, knowledge, and harmony, the QDC framework suggests a sublinear scaling law between computing resources and quantum simulations, minimizing computational complexity when compared with a linear scaling law due to Feynman's suggestion of direct proportionality.

The quantum computing industry today has largely adhered to Richard Feynman's vision of a probabilistic quantum computer [1]. This vision is rooted in the Copenhagen Interpretation of Quantum Mechanics [2], which posits that it is fundamentally impossible to fully understand the quantum realm. In this framework, quantum dynamics are defined by the collapse of potential states into measurable outcomes during observation, making the quantum domain dependent on the act and method of measurement.

In contrast, the Qualified Determinism Construct (QDC) model presented in this treatise proposes an alternative perspective on quantum-level behavior[3].[1] The model introduces an intrinsic fourfold structure within quantum space, which continually manifests in infinitely varied but ordered ways. This perspective shifts the

[1] An initial IEEE article [3] focused on building on Feynman's conceptions of simulating time, space, probabilities, and estimating the size of quantum computers. In this chapter these conceptions are reframed to be in line with the overall model in the book.

P. Malik, *Pioneering New Avenues in Quantum Technology*, Studies in Smart Technologies, https://doi.org/10.1007/978-981-96-5463-5_9

understanding of nature beyond the confines of statistical and probabilistic interpretations. Consequently, while Feynman's probabilistic quantum computer seeks to simulate space, time, and probability, the QDC framework enhances this by reimagining space and time as emergent properties governed by precise meta-functional logic and replacing the lens of probability with one of structured fourfoldness. This paradigm sets the stage for a fundamentally new genre of quantum computers.

This new genre gains further credibility by reinterpreting foundational quantum equations such as Schrödinger's Wave Equation (Chap. 7), Heisenberg's Uncertainty Principle (Chap. 5), and Euler's identity (Chap. 8). These reinterpretations redefine the nature of quantum computation as a creative process, rather than merely a constructive one.

Richard Feynman's fundamental perspective on quantum computation emphasizes the ability to simulate nature. The complexity of such simulation depends on one's interpretation of how nature operates. Feynman's own approach is rooted in the Copenhagen Interpretation of Quantum Mechanics, which asserts that the true nature of quantum phenomena cannot be fully understood. Instead, measurements are performed, leading to the collapse of the infinite possibilities (or superpositions) within a quantum object into observable states. A representation of quantum dynamics is then constructed from repeated measurements, where statistically significant patterns are used as proxies for the processes occurring within the quantum "black box."

In his seminal paper "Simulating Physics with Computers," Feynman [1] explains this approach by suggesting that a probabilistic nature, denoted as N, could potentially be simulated using a computer C that is also probabilistic. In this approach, the last two digits of every number could be randomized, or another form of stochastic perturbation could be introduced, effectively making C a probabilistic computer. In such a system, the output would no longer be a unique function of the input. The goal would then be to configure C in a way that its transitions from an initial state to a final state occur with the same probability as the corresponding transitions in N. This alignment would allow C to effectively simulate the probabilistic behavior of nature. Feynman acknowledges that such a probabilistic computer does not precisely replicate nature's processes but rather matches its probabilities. He suggests that nature is inherently unpredictable, making it impossible to predict its behavior with absolute certainty using a computer, particularly if it follows probabilistic laws. However, in a probabilistic system, repeating an experiment in nature multiple times allows for an approximation of its behavior based on observed frequencies. Similarly, if a computer repeats the same experiment numerous times, it can generate outcomes with frequencies proportional to those observed in nature. While the machine does not precisely replicate nature's processes, it can serve as a probabilistic simulator. By running repeated simulations, the computer can approximate nature's probabilities with a comparable level of accuracy, providing a statistically reliable representation of its probabilistic behavior.

This probabilistic paradigm underpins quantum computers developed by industry leaders like Google and IBM [4]. However, there remains a critical limitation in this perspective: if nature were inherently random, it would fail to produce and sustain the intricate and repeatable structures observed in matter and life. Such structures would be subject to continuous quantum impulses, causing them to evolve unpredictably into entirely different forms over time. Evidently, this is not the case, suggesting that nature's underlying processes extend beyond pure randomness, requiring deeper exploration into deterministic and emergent principles.

9.1 Simulating Space

Richard Feynman posits the concept of a probabilistic quantum computer that, unlike classical digital computers, could simulate probabilities by leveraging quantum elements. If such a computer were to analyze phenomena within a small space–time volume, it would require only a finite number of operations. However, Feynman acknowledges that this is not how physics traditionally operates. Physics allows space to exist at infinitesimal scales, whereas a computer simulating physics would need to treat space as composed of discrete and finite elements [5], aligning with the finite operations required for computation. Interestingly, decades after Feynman's suggestion, modern physics is converging on the idea that space exhibits granularity [6].

This perspective aligns with the Qualified Determinism Construct (QDC), where space is seen as a field hosting "seeds" formed when light precipitates to the speed of c. If space is granular in this way, simulating it on a quantum computer involves more than mere discretization—it requires recognizing a governing meta-function. Moreover, a complete simulation must incorporate the interconnected dynamics of the space–time-energy-gravity quadrumvirate.

The significance of this insight lies in Feynman's foundational stance on quantum computing: the quantum computer must accurately simulate nature. However, if space inherently contains archetypes of knowledge, the current quantum computing industry, rooted in a statistical-probabilistic interpretation of nature, overlooks a critical element. This limited lens misses the deeper structures and meta-functions inherent in nature's dynamics, thus constraining the industry's ability to truly replicate the processes of the natural world.

9.2 Size of Quantum Computer

Feynman proposed that the size of a quantum computer should be directly proportional to the space–time volume of the physical system it aims to simulate. This correlation arises from the view that the quantum layer consists of a single fundamental layer, from which all emergent behaviors arise unpredictably. Under this paradigm, the unpredictable nature of quantum phenomena necessitates a direct scaling relationship between the computer's size and the space–time volume being simulated.

However, if we adopt a symmetrical, multi-layered UQO model in which emergent phenomena are projections of the foundational four properties—space, time, energy, and gravity—then the simulation is governed by a hierarchy of meta-functions. These meta-functions determine the behavior of space–time-energy-gravity interactions, reducing the computational complexity of simulations. In this context, the direct proportionality suggested by Feynman can be replaced by a sublinear scaling law, where the size of the quantum computer increases only incrementally, if at all, as the space–time volume grows. This is because the multi-layered model leverages the structured and deterministic interactions of meta-functions, minimizing the computational overhead required to simulate increasingly large systems.

9.3 Simulating Time

In the envisioned quantum computer, time, like space, must be discretized. Using cellular automata, it may become feasible to imitate time by breaking it into discrete intervals. However, Feynman raises the critical question of whether time can be simulated rather than merely imitated. Achieving true simulation would give the quantum computer a deeper, lens-like capability to reveal the dynamics of the quantum realm. Feynman suggests that points in space are not only influenced by other points in space at the same moment but also potentially by points in space at a future time, incremented by one or more discrete intervals.

Feynman's approach to simulating time, like his approach to simulating space, is fundamentally linear. However, as suggested by the Qualified Determinism Construct (QDC), the universe operates more as a complex adaptive system, better captured by the holographic, fractal QDC equation previously outlined (4.15).

In the QDC framework, time is emergent and relative. It reflects the maturation of a meta-function governing a seed in space, which evolves at its own pace, influenced by real-time interactions within a given space–time volume. While neighboring points in space may exert an influence on the dynamics of a seed, the meta-function specific to that seed plays the most critical role in shaping how time is experienced.

To simulate time in a quantum computer using QDC principles, it is necessary to move beyond linear assumptions and adopt the multi-layered, emergent perspective defined by the QDC model. This requires modeling time not as a static, universal dimension but as a dynamic, seed-specific property shaped by meta-layer influences and interactions within the space–time-energy-gravity framework.

9.4 Simulating Probabilities

Feynman, expanding on the prevailing understanding of quantum dynamics, suggests that quantum mechanics inherently involves probability, making it essential to consider how probability can be simulated. One approach to simulating a probabilistic theory on a computer is to calculate the relevant probabilities and then interpret these values as representations of nature's behavior. This method allows the computer to approximate the inherent uncertainty and randomness found in quantum systems.

He highlights the significant challenge of simulating systems with many particles, which requires exponentially increasing computational resources to calculate probabilities in larger spaces. This reasoning stems from a fundamentally linear framework and assumes that nature operates primarily on probabilistic principles. While probability is undoubtedly a useful tool for summarizing phenomena in an unfathomable realm, combining it with statistical analysis—where repeated measurements yield observable trends—provides a comforting but limited understanding of quantum dynamics. This probabilistic interpretation, as Feynman suggests, underpins the conceptual foundation of such a system.

However, when dealing with the quantum realm—a domain that bridges the invisible and the visible—relying solely on observable phenomena without constructing models to explain the underlying principles offers limited insight. Unfortunately, this is precisely the approach the quantum computing industry has largely adopted. Following Feynman's description, industry leaders have focused on building probabilistic quantum computers. These machines are designed to simulate nature by ensuring their transitions from initial to final states align probabilistically with nature's transitions, as defined by corresponding initial and final states in a computational context. The output of such computers is inherently non-deterministic, as the machines are probabilistic by design.

This approach forms the cornerstone of today's quantum computing industry. Yet, it ultimately falls short of addressing the deeper quantum dynamics that drive the emergence of matter, life, and other phenomena. This is where models like the Qualified Determinism Construct (QDC) may offer a more valuable framework. By shifting the focus toward experimental designs and assessment tools aligned with QDC principles, we may gain a more profound understanding of the mechanisms underpinning quantum-level emergences, moving beyond the constraints of probabilistic paradigms.

9.5 Further Reading

- "Feynman Lectures on Computation" edited by Anthony J.G. Hey and Robin W. Allen: A compilation of Feynman's lectures on computation, including his ideas on quantum computation.

- "Causality: Models, Reasoning and Inference" by Judea Pearl: A comprehensive treatment of causality and its implications for various fields, including science and artificial intelligence. This can provide a framework for thinking about deterministic and probabilistic explanations.
- "Foundations of Higher-Order Logic" by Peter B. Andrews: A more advanced and specialized book focusing specifically on the foundations and properties of higher-order logic.
- "Category Theory in Context" by Emily Riehl: An accessible introduction to category theory, with a focus on applications in various areas of mathematics.
- "Topos Theory" by Colin McLarty: Topos theory is a more advanced area of category theory that has found applications in logic, set theory, and even theoretical physics. It provides a powerful framework for studying different kinds of mathematical structures and their relationships.

References

1. Feynman RP (1982) Simulating physics with computers. Int J Theor Phys 21(6):467–488
2. Faye J (2024) Copenhagen interpretation of quantum mechanics. In: Stanford encyclopedia of philosophy. Stanford University, 31 May 2024. https://plato.stanford.edu/entries/qm-copenhagen/
3. Malik P (2022) Enhancing Feynman's quantum computational positioning to inject new possibility into the foundations of the quantum computing industry. In: 2022 IEEE 13th annual information technology, electronics and mobile communication conference (IEMCON), Vancouver, BC, Canada, 2022, pp 0546–0553. https://doi.org/10.1109/IEMCON56893.2022.9946451
4. What is quantum computing? IBM Think. Accessed 13 Jan 2025. https://www.ibm.com/think/topics/quantum-computing
5. Feynman RP (2017) Seeking new laws. In: The character of physical law: the 1964 messenger lectures. MIT Press, Cambridge (2017)
6. Granular space and the problem of large numbers. J Mod Phys 3(8):690–695 (2012). https://www.scirp.org/html/4869.html

Chapter 10
Simulation of a Hypothetical Simple Atom

Abstract This chapter presents a simulation framework for modeling a hypothetical atom using Qualified Deterministic Construct (QDC) and Quantum Object Code (QOC), focusing on the atom's quantized evolution across space, time, energy, and gravity. Four key evolution equations, derived from QOC principles, form the basis of a higher-dimensional 4D wavefunction. A modified Schrödinger equation governs the system's dynamic behavior, integrating intrinsic seed behaviors and inter-seed interaction dynamics, which are modeled using distance-dependent potentials and coupling constants. The simulation process involves defining seed properties, assigning baseline values, computing evolution functions, and iteratively updating seed properties based on net interaction forces, ultimately extracting emergent patterns and behaviors from the evolving 4D wavefunction. This approach provides a robust framework for simulating atom-like systems, emphasizing the interplay between fundamental properties, quantization, interaction dynamics, and the emergence of complex quantum phenomena.

To give a sense of how simulation of a hypothetical simple atom would be set up using UQO, QDC, and Quantum Object Code (QOC) here some governing equations, behavioral dynamics and interaction models, a modified Schrodinger equation, and simplified steps to set up such a simulation will be looked at.

10.1 Key Governing Equations

A seed for a hypothetical atom is illustrated by (10.1) and is representative of M1 dynamics in UQO (4.5) and QDC (4.15):

$$\text{Seed} = Xa + \sum_{i=0}^{n} Y_i b_i \quad \text{where} \begin{bmatrix} X, Y \in [S_{Pr}, S_{Po}, S_K, S_H] \\ a, b \text{ are integers; } a > b \end{bmatrix}. \tag{10.1}$$

P. Malik, *Pioneering New Avenues in Quantum Technology*, Studies in Smart Technologies, https://doi.org/10.1007/978-981-96-5463-5_10

The key governing equations for simulation of a hypothetical atom derive from the QOC Eqs. (6.3)–(6.8).

Equation (10.2), *Space Evolution*, is a vector equivalent of (6.3) for space quantization and assumes that the seed represented by (4.8) has been quantized and is now available at B as specified by UQO and QDC.

$$S_e = h\left(\vec{X} + \sum_{i=0}^{n} \vec{Y}_i\right).\Theta(\text{seed}). \tag{10.2}$$

Here, $X, Y \in [S_{Pr}, S_{Po}, S_K, S_H]$; $n = 3$ (a simple atom with three secondary properties); $\|\vec{X}\| > \|\vec{Y}_i\|$; h—Planck's constant; and $\Theta(\text{seed})$ is a Heaviside function that acts as a logical gate to ensure that a valid seed is in place.

Equation (10.3), *Temporal Evolution*, derives from the equation for time quantization (6.5):

$$T_e = h\left(T_o + \tilde{\vec{X}} + \sum_{i=0}^{n} \tilde{\vec{Y}}_i\right).\Theta(S_e). \tag{10.3}$$

Here

- T_o-baseline temporal evolution state/value.
- $X, Y \in [S_{Pr}, S_{Po}, S_K, S_H]$.
- $I_{S_{Po}}\left(\vec{X}, \vec{Y}\right) = 1$ if X or Y is an element (I) of S_{Po} (set of Power).
- $\tilde{\vec{X}}$—adapted primary property for temporal adaptability, derived from the X-term in S_q, such that $\tilde{\vec{X}} = 1 - 0.2\left(2.I_{S_{Po}}\left(\vec{X}\right) - 1\right)$ implying that time to maturity is reduced by 20% if $I_{S_{Po}}\left(\vec{X}\right) = 1$, else it is increased by 20% (this follows the connection between 'time' and 'power' established in Chap. 2).
- $\tilde{\vec{Y}}_i$—adapted primary property for temporal adaptability, derived from the Y_i-term in S_q, such that $\tilde{\vec{Y}}_i = 1 - 0.02\left(2.I_{S_{Po}}\left(\vec{Y}_i\right) - 1\right)$ implying that time to maturity is reduced by 2% if $I_{S_{Po}}\left(\vec{Y}_i\right) = 1$, else it is increased by 2%.
- $\|\vec{X}\| > \|\vec{Y}_i\|$.
- Θ—Heaviside function that acts as a logical gate, since time quantization will act on the same seed that has been quantized into space implying that $\Theta(S_e) = 1$.
- h—Planck's constant.

Equation (10.4), *Energy Evolution*, derives from the equation for energy quantization (6.6).

$$E_e = h\left(E_o + \tilde{\vec{X}} + \sum_{i=0}^{n} \tilde{\vec{Y}}_i\right).\Theta(S_e).(1 + \eta.\mathcal{M}(S_k)), \tag{10.4}$$

where

- E_o—represents the intrinsic energy of the seed.
- $X, Y \in [S_{Pr}, S_{Po}, S_K, S_H]$.
- $I_{S_{Pr}}\left(\vec{X}, \vec{Y}\right)$—is $= 1$ if X or Y is an element (I) of S_{Pr} (set of presence).
- $\tilde{\vec{X}}$—adapted primary property for energy evolution, derived from the X-term in S_q, such that $\tilde{\vec{X}} = 1 + 0.2\left(2.I_{S_{Pr}}\left(\vec{X}\right) - 1\right)$ implying that energy is augmented by 20% if $I_{S_{Pr}}\left(\vec{X}\right) = 1$, else it is retarded by 20% (this follows the connection between 'energy' and 'presence' established in Chap. 2).
- $\tilde{\vec{Y}}_i$—adapted primary property for energy evolution, derived from the Y_i-term in S_q, such that $\tilde{\vec{Y}}_i = 1 + 0.02\left(I_{S_{Pr}}\left(\vec{Y}_i\right) - 1\right)$ implying that energy is augmented by 2% if $I_{S_{Pr}}\left(\vec{Y}_i\right) = 1$, else it is reduced by 2%.
- $\|\vec{X}\| > \|\vec{Y}_i\|$.
- $\Theta(S_e)$—signifies spatial dependency such that energy quantization will occur only on a seed that has already been quantized for space (S_e).
- $\mathcal{M}(S_k) = \sum_{k \in \{Pr, Po, K, H\}} I\left(\left|\left(\vec{X} \cup \left\{\vec{Y}_i\right\}\right) \cap S_k\right| > 1\right)$, where $\left(\vec{X} \cup \left\{\vec{Y}_i\right\}\right) \cap S_k$ represents the properties that belong to S_k. I is Indicator function which equals 1 if the set contains more than 1 property. $\mathcal{M}(S_k)$ therefore counts the sets with more than one property.
- η—Acceleration function of 10%, implying energy available toward materialization increased by 10% with every additional set that has 2 or more properties.
- h—Planck's constant.

Equation (10.5), *Gravity Evolution*, derives from the equation for gravity quantization (6.7):

$$G_e = h\left(G_o + \tilde{\vec{X}} + \sum_{i=0}^{n} \tilde{\vec{Y}}_i\right).\Theta(S_e).\left(1 + \eta.\mathcal{U}\left(\vec{X}, \left\{\vec{Y}_i\right\}\right)\right), \tag{10.5}$$

where

- G_o—represents baseline gravitational behavior of the seed.
- $X, Y \in [S_{Pr}, S_{Po}, S_K, S_H]$.
- $I_{S_H}\left(\vec{X}, \vec{Y}\right)$—is $= 1$ if X or Y is an element (I) of S_H (set of Harmony).
- $\tilde{\vec{X}}$—adapted primary property for gravity evolution, derived from the X-term in S_q, such that $\tilde{\vec{X}} = 1 + 0.2\left(2.I_{S_H}\left(\vec{X}\right) - 1\right)$ implying that conditions for harmony are augmented by 20% if $I_{S_H}\left(\vec{X}\right) = 1$, else it is retarded by 20% (this follows the connection between 'gravity' and 'harmony' established in Chap. 2).

- $\tilde{\vec{Y}}_i$—adapted primary property for gravity evolution, derived from the Y_i-term in S_q, such that $\tilde{\vec{Y}}_i = 1+0.02\left(2.I_{S_H}\left(\vec{Y}_i\right) - 1\right)$ implying that conditions for harmony are augmented by 2% if $I_{S_H}\left(\vec{Y}_i\right) = 1$, else it is retarded by 2%.
- $\|\vec{X}\| > \|\vec{Y}_i\|$.
- $\Theta(S_e)$—signifies spatial dependency such that gravity evolution will occur only on a seed that has already been quantized for space (S_e).
- $\mathcal{U} = 1$: Seed spans all four sets; Hence, $\mathcal{U}\left(\vec{X}, \left\{\vec{Y}_i\right\}\right) = 1$ if $\mathcal{U}\left(\vec{X} \cup \left\{\vec{Y}_i\right\}\right) \in S_{Po} \cap S_{Pr} \cap S_K \cap S_H$; 0, otherwise.
- η—Acceleration function of 20%, implying that if $\mathcal{U}\left(\vec{X}, \left\{\vec{Y}_i\right\}\right) =1$ then conditions for harmony at the level of particles/atoms are more ripe by an additional 20%.
- h—Planck's constant.

In Eq. (10.6), *Higher-Dimensional Wavefunction*, these four evolution equations (right-hand side) can be thought of as components of a higher-dimensional 4D wavefunction (left-hand side), with each dimension having infinite degrees of freedom, represented by:

$$\psi(S, T, E, G) = \psi(S_e, T_e, E_e, G_e). \tag{10.6}$$

Therefore (10.6) reflects the system's specific quantized states when evaluated at S_e, T_e, E_e, G_e.

10.2 A Seed's Behavior

A seed's behavior, where behavioral dynamics refer to how a single seed behaves within the 4D quantum space [1] as influenced by its equations for evolution, can be expressed by Eq. (10.7), *Unified Behavioral Function*, which specifies the i-th seed's behavior:

$$F_{B,i} = \Phi\left(S_{e,i}, T_{e,i}, E_{e,i}, G_{e,i}\right). \tag{10.7}$$

Here, Φ is a meta-function combining contributions from each energy Eq. (10.2)–(10.5).

The interaction potential between seeds is given by Eq. (10.8), *Interaction Potential*:

$$F_{\text{interaction}}(i, j) = \psi\left(r_{ij}\right).\Gamma\left(\overrightarrow{F_{B,i}}, \overrightarrow{F_{B,j}}\right). \tag{10.8}$$

Here

- r_{ij}—distance between seeds i, j.
- $\psi\left(r_{ij}\right) = \frac{1}{r_{ij}^{n}}$—distance-dependent interaction strength.
- $\Gamma\left(\overrightarrow{F_{B,i}}, \overrightarrow{F_{B,j}}\right) = \kappa \prod_{\upsilon \varepsilon \{S,T,E,G\}} \left(\upsilon_{i,e} . \upsilon_{j,e}\right)$—Behavioral coupling between seeds i and j, considering contributions from all dimensions; κ—coupling constant.

The total interaction force acting on seed i is depicted by Eq. (10.9), *Net Interaction Force*:

$$F_{\text{Net, interaction},i} = \sum_{j \neq i} F_{\text{interaction}}(i, j). \tag{10.9}$$

10.3 Modified Schrodinger's Equation

A modified Schrodinger equation representing the dynamic evolution of the 4D quantum system, incorporating both the intrinsic behavior of individual seeds and their interactions is depicted by Eq. (10.10), *Modified Schrodinger Equation*:

$$i\hbar \frac{\partial}{\partial t} \psi(S, T, E, G) = \hat{H} \psi(S, T, E, G). \tag{10.10}$$

Here

- $\psi(S, T, E, G) = \psi(S_e, T_e, E_e, G_e)$—this is the 4D wavefunction implying that the 4D wavefunction encapsulating the quantum system as a single entity can be evaluated at specific quantized states $\psi(S_e, T_e, E_e, G_e)$.
- $\hat{H} = \hat{H}_S + \hat{H}_T + \hat{H}_E + \hat{H}_G$—Hamiltonian operator with contributions from space, time, energy, gravity dynamics. Each component of $\hat{H}$ derives from the corresponding energy Eqs. (10.2)–(10.5), accounting for the seed's evolving properties. $\hat{H}$ also accounts for interconnected dynamics between seeds.

10.4 Simulation Process

The simulation process will have the following steps:

1. Define the seed's functional properties specifying X (primary factor) and Y (secondary factors): This step identifies the key characteristics of the atom or particle being simulated. For example, the primary property (X) could represent a dominant feature such as malleability or reactivity, while secondary properties (Y) could represent supplementary traits like conductivity or ductility. These properties determine how the seed will behave in the quantum space.

2. Assign baseline values—T_o, E_o, G_o: Each seed is given initial values for its time evolution, energy state, and gravitational interaction. These act as starting points for the simulation and provide a foundation for quantization.
3. Assign acceleration function η for specific sets S_{Po}, S_{Pr}, S_H: The acceleration function is used to modify how quickly certain behaviors evolve. Specific sets (e.g., S_{Po} for power, S_{Pr} for presence, S_H for harmony) determine if acceleration is applied to a seed's temporal, energy, or gravitational dynamics, making those processes occur faster.
4. Compute evolution functions—S_e, T_e, E_e, G_e: The simulation calculates how each seed evolves in space, time, energy, and gravity. These calculations are influenced by the functional properties, baseline values, weights, and acceleration factors. The results describe the seed's state across all dimensions at this stage of the simulation.
5. Compute behavioral dynamics—$F_{B,i}$: Behavioral dynamics represent how a seed behaves as a result of its quantized properties. This step combines the outputs from the evolution functions to define the seed's overall "personality" or response in the quantum space.
6. Compute interaction dynamics for each pair of seeds—distance-dependent potential $\left(\psi\left(r_{ij}\right)\right)$, behavioral coupling $\left(\Gamma\left(F_{B,i}, F_{B,j}\right)\right)$, interaction force $(F_{\text{interaction}}(i, j))$: Interaction dynamics determine how seeds influence each other. This involves distance-dependent potential, how strongly two seeds interact based on their distance; behavioral coupling—how the behaviors of two seeds affect one another; and, interaction force—the total influence one seed exerts on another, considering both distance and behavioral effects.
7. Update properties of each seed based on interactions, adjusting S_e, T_e, E_e, G_e based on $F_{\text{Net, interaction},i}$: The properties of each seed (spatial, temporal, energy, gravitational) are adjusted based on the forces resulting from interactions with other seeds. This step ensures that the simulation accounts for dynamic changes as seeds interact over time.
8. Iterate the simulation over discrete time steps using the modified Schrodinger equation, $i\hbar\frac{\partial}{\partial t}\psi(S, T, E, G) = \hat{H}\psi(S, T, E, G)$: The simulation progresses in discrete time steps. At each step, the system's state is updated based on a modified Schrödinger equation, which governs how the quantum wavefunction evolves. This iterative process allows the simulation to capture continuous changes.
9. Extract emergent properties from $\psi(S, T, E, G)$: The simulation analyzes the combined quantum wavefunction to identify emergent features. These might include patterns or behaviors that arise from the collective interaction of seeds, such as stability, coherence, or new structures.

10.5 Further Reading

- "Computational Physics" by Nicholas J. Giordano and Hisao Nakanishi: A good introduction to numerical methods used in physics simulations.

- "Numerical Recipes: The Art of Scientific Computing" by William H. Press et al.: A comprehensive guide to numerical algorithms and techniques.
- "Classical Mechanics" by Herbert Goldstein, Charles Poole, and John Safko: A graduate-level text on classical mechanics, covering Lagrangian and Hamiltonian mechanics, which are relevant to the concept of evolution functions.
- "Agent-Based and Individual-Based Modeling: A Practical Introduction" by Steven F. Railsback and Volker Grimm: A practical guide to agent-based modeling, which is relevant to the concept of interacting seeds.
- "Solving Ordinary Differential Equations I: Nonstiff Problems" by Ernst Hairer, Syvert P. Nørsett, and Gerhard Wanner: A standard reference on numerical methods for solving ordinary differential equations, which are used to model time evolution.

Reference

1. Malik P (2022) The role of a light-based quantum computational model in the creation of an oscillating universe. In: 2022 IEEE 12th annual computing and communication workshop and conference (CCWC), Las Vegas, NV, USA, 2022, pp 0953–0959. https://doi.org/10.1109/CCWC54503.2022.9720904

Chapter 11
Quantum Certainty

Abstract This chapter explores the cosmos as a complex adaptive system (CAS) where the interplay of space, time, energy, and gravity (STEG) properties drives emergent phenomena. It proposes a model where persistent quantum computation, coupled with feedback mechanisms, guides adaptive changes through a process called "quantum certainty." This is illustrated by an insect adapting to its environment: a need for protection triggers access to deeper organizational layers, generating a new "protective shell" meta-function. This adaptation unfolds through coordinated STEG quantization where space encodes new information, time structures the change, energy materializes the shell, and gravity adjusts environmental interactions. This framework highlights the transformative potential of CAS, where quantum phenomena, driven by a deep urge for change, alter system dynamics, with implications for both biological adaptation and quantum computation.

The quantum model and the accompanying architecture and algorithms proposed in this book are rooted in a fractal model of the cosmos, as outlined in the 20-volume *Cosmology of Light* book series [1]. This cosmology emerges from imagining various speeds of light, each giving rise to distinct realities. Chapter 2 elaborates on this framework, introducing four Hilbert spaces (2.1)–(2.4) to encapsulate these realities. When these realities are interwoven, a comprehensive cosmology arises in which materialization is understood as the interplay of these interconnected realities. The mathematical framework initially proposed in the *Cosmology of Light* series is further refined and extended in this book, employing established quantum mathematics and introducing quantum computing architectures to operationalize this vision.

The cosmos is conceptualized as a complex adaptive system (CAS), inherently possessing mechanisms that enable emergent phenomena to influence fundamental dynamics. This chapter links such dynamics to UQO (11.1) and QDC (11.2) through the lenses of adaptation and genetics.

A model integrating feedback mechanisms is proposed in this chapter, connecting the foundational reality of persistent quantum computation to outputs resembling genetic information [2]. This genetic-type information can modify space–time-energy-gravity (STEG) dynamics based on emergent properties, informing the design

P. Malik, *Pioneering New Avenues in Quantum Technology*, Studies in Smart Technologies, https://doi.org/10.1007/978-981-96-5463-5_11

of alternative quantum computation hardware and algorithms. To address the concept of quantum uncertainty in a qualified deterministic model as surfaced in Chap. 5, this chapter introduces the notion of quantum certainty, achieved through feedback loops that enable specific fourfold quantization processes. This quantum certainty can be initiated by a profound urge or will for transformation [3–6].[1]

A practical illustration of this framework is provided through insect adaptation, where STEG quantization serves as a mechanism encoding genetic adaptation in a coordinated fourfold script. This composite script facilitates the creation or invocation of a fourfold quantization meta-function capable of inducing material-level changes. Furthermore, the integration of feedback mechanisms involving living systems, quantum certainty, and STEG quantization becomes pivotal in realizing the quantum computing possibilities associated with the $u(W(L))$, $v(W(L))$, and $w(L)$ formulations introduced in Chap. 1.

11.1 Quantum Certainty

By drawing on the existing quantum mathematical model, this chapter examines the growth of fourfold complexity through a representative case: insect adaptation. As summarized by the Unified Quantum Object (UQO) in (11.1)—reproduced below—the quantum level provides a vantage point into layers of reality that lie beyond the apparent surface described by $|\psi_c\rangle$. This view is reinforced by earlier interpretations: Schrödinger's equation (implying matter as a product of superposed states with boundless possibilities), Heisenberg's uncertainty principle (describing constant quantum fluctuations from an ever-present "pregnant-infinity"), and Euler's identity [depicting materialization as a cumulative process across antecedent layers outlined in (11.2)].

$$\mathrm{UQO} = \begin{bmatrix} |\psi_\infty\rangle = m(\mathrm{Pr}, \mathrm{Po}, K, H) \\ (\downarrow P_k|\psi_\infty\rangle) \\ |\psi_k\rangle = n(S_{\mathrm{Pr}}, S_{\mathrm{Po}}, S_K, S_H) \\ (\downarrow P_g|\psi_k\rangle) \\ |\psi_g\rangle = o(S_{\mathrm{Pr}} x S_{\mathrm{Po}} x S_K x S_H) \\ (\downarrow P_c|\psi_g\rangle) \\ |\psi_c\rangle = p(S, T, E, G) \end{bmatrix}. \tag{11.1}$$

Any system whose outcomes hinge on interactions across these layers also requires feedback processes that exploit loops to propagate effects. Although the visible layer $|\psi_c\rangle$ is always shaped by deeper strata, quantum computation may unfold differently

[1] Note that the Cosmology of Light book series [1] and in particular the book on Quantum Certainty [4], and further IEEE article [3] lay out the initial thought and early mathematical formulation of adaptation through 'quantum certainty'. In this chapter, the initial math has been developed to be consistent with the novel mathematical framework being developed in this book.

if activity in the B-Layer ($|\psi_c\rangle$) is established by any of the meta-layers—$M1$ ($|\psi_g\rangle$), $M2$ ($|\psi_k\rangle$), or $M3$ ($|\psi_\infty\rangle$). It is possible to construct a computational state that selectively engages "meta-functions" or specialized seeds residing in $|\psi_g\rangle$. Once initiated, these states can trigger processes that achieve quantum certainty and yield fourfold STEG quantization, potentially modifying the underlying "law" or micro-script.

These activation conditions are described by QDC (11.2), classifying B, $M1$, $M2$, and $M3$ as successive levels within UQO:

$$\text{QDC} = \begin{bmatrix} M3 \to m(\text{Pr}, \text{Po}, K, H) \\ \downarrow\uparrow \\ M2 \to n(S_{\text{Pr}}, S_{\text{Po}}, S_K, S_H) \\ \downarrow\uparrow \\ M1 \to o(S_{\text{Pr}} \times S_{\text{Po}} \times S_K \times S_H) \\ \downarrow\uparrow \\ B \to p(S, T, E, G) \end{bmatrix}. \tag{11.2}$$

In essence, if the computational system at level B is "overcome," the capabilities residing at successively higher layers ($M1$, $M2$, $M3$) are unlocked, provided hardware, software, or algorithmic conditions are revised at B. For instance, envision a hypothetical insect—functioning at B—that is continually targeted by predators. A powerful urge to withstand some predators may arise, effectively breaking long-held species patterns. Similar discussions exist on insect behavior in fragmented habitats and computational models of insect cognition [7, 8]. If this impetus is sufficiently strong, it can act as an activation state that opens a quantum window, enabling interaction with $M1$, or even $M2$ or $M3$, depending on how profound the urge is.

This impulse can foster an exchange with a "collective-intelligence meta-function" or "specific-species-intelligence meta-function," originating from a particular seed as noted in [3]. Through this route, a "predator-overcoming meta-function" can emerge, allowing the insect to adapt to avoid at least certain predators—perhaps by evolving a sturdy shell. From a mathematical standpoint, and drawing on the space quantization (6.3), such a meta-function might be formulated by Eq. (11.3), *Hard Protective Shell Meta-Function*:

$$\text{Sig}_{\text{hard-protective-shell}} = \left(a\vec{X} + \sum_{i=0}^{n} b_i \vec{Y}_i\right).\Theta(\mathcal{A}(M1, M2, M3)). \tag{11.3}$$

Here

- $\vec{X} \in [S_{\text{Po}}]$.
- $\overrightarrow{Y_i} \in [S_{\text{Pr}}, S_{\text{Po}}, S_K, S_H]$.
- a and b are integers such that $a > b$.
- $\Theta(\mathcal{A}(M1, M2, M3))$: Activation condition with Heaviside-like function, Θ, to ensure transformation occurs when at least one of $M1$, $M2$, $M2$ is active. $\mathcal{A}(M1, M2, M3)$ is the activation state reflecting meta-layer interactions.

In this context, S_{Po} points to polysaccharides—linked to the property of power—which adapt to form a protective carapace. Thus, X is drawn from the set of polysaccharides. Meanwhile, S_{Pr} relates to presence (proteins), S_K to knowledge (nucleic acids), and S_H to harmony (lipids). The secondary element '$\sum_{i=0}^{n} b_i Y_i$' denotes potentially multiple contributions, including proteins, polysaccharides, nucleic acids, and lipids, all working in tandem to manifest $\mathrm{Sig}_{\text{hard-protective-shell}}$. For instance, nucleic acids might encode instructions for how proteins, in combination with existing polysaccharides and lipids, assemble into a new protective shell. Eventually, specialized cells arise to reinforce the insect's defenses.

11.2 Space–Time-Gravity-Energy or STEG Quantization

Although the disruption of existing patterns is an initial phase, and generating a novel meta-function is a second, the third phase initiates specific quantization, thus completing a feedback loop within the universal CAS. This progression can be described using a variation of (11.3), shown in Eq. (11.4), *Generalized Seed Change at Meta-Function Level*:

$$\mathrm{Seed}_{\mathrm{mfc}} = \left(a\vec{X} + \sum_{i=0}^{n} b_i \vec{Y}_i\right).\Theta(\mathcal{A}(M1, M2, M3)). \tag{11.4}$$

Here, $\vec{X} \in [S_{\mathrm{Po}}]$; $\vec{Y_i} \in [S_{\mathrm{Pr}}, S_{\mathrm{Po}}, S_K, S_H]$; a and b are integers such that $a > b$; 'mfc'—meta-function-change.

According to (11.3), the emergence of a hard-protective-shell trait results from closely integrated quantization of space, time, energy, and gravity (STEG). Equations (11.5)–(11.8) specify how these quantizations operate as mechanics of adaptation, showcased here through an insect example.

Hence, there is a quantization as specified by Eq. (11.5), *Space-Quantization* that leverages off (6.3):

$$\mathrm{Space}_{\mathrm{quantization}} = h\left(aX + \sum_{i=0}^{n} b_i Y_i\right).\Theta(S_{\mathrm{mfc}}). \tag{11.5}$$

Here, $X \in [S_K]$; $Y \in [S_{\mathrm{Pr}}, S_{\mathrm{Po}}, S_K, S_H]$; a and b are integers such that $a > b$; h—is Planck's constant; $\Theta(S_{\mathrm{mfc}})$ captures (11.4) suggesting that the seed needs to change before quantization takes place.

Through (11.4), space is reconfigured so that new "knowledge seeds" associated with the meta-function become woven into the STEG-Fabric, enabling the species to more effectively realize its adaptive changes.

Further, there is a *time* quantization leveraging off (6.5) as specified by Eq. (11.6), *Time Quantization*:

$$\text{Time}_{\text{quantization}} = h\left(aX + \sum_{i=0}^{n} b_i Y_i\right).\Theta(S_{\text{mfc}}). \tag{11.6}$$

Here, $X \in [S_{\text{Po}}]$; $Y \in [S_{\text{Pr}}, S_{\text{Po}}, S_K, S_H]$; a and b are integers such that $a > b$; h—is Planck's constant; $\Theta(S_{\text{mfc}})$ captures (11.4) suggesting that the seed needs to change before quantization takes place.

Time quantization encodes phases of growth into temporal structures, enabling varied rates of change aligned with the species' pursuit of predator evasion.

There is also an *energy* quantization leveraging off (6.6), specified by Eq. (11.7), *Energy Quantization*:

$$\text{Energy}_{\text{quantization}} = h\left(aX + \sum_{i=0}^{n} b_i Y_i\right).\Theta(S_{\text{mfc}}). \tag{11.7}$$

Here, $X \in [S_{\text{Pr}}]$; $Y \in [S_{\text{Pr}}, S_{\text{Po}}, S_K, S_H]$; a and b are integers such that $a > b$; h—is Planck's constant; $\Theta(S_{\text{mfc}})$ captures (11.4) suggesting that the seed needs to change before quantization takes place.

Energy quantization (11.7) allows *matter* to be formed, and in this case, will be tied to energy quantization, supporting incremental physical modifications like forming a hardened shell.

Finally, there is envisioned to be a *gravity* quantization leveraging off (6.7), specified by Eq. (11.8), *Gravity-Quantization*:

$$\text{Gravity}_{\text{quantization}} = h\left(aX + \sum_{i=0}^{n} b_i Y_i\right).\Theta(S_{\text{mfc}}). \tag{11.8}$$

Here, $X \in [S_H]$; $Y \in [S_{\text{Pr}}, S_{\text{Po}}, S_K, S_H]$; a and b are integers such that $a > b$; h—is Planck's constant; $\Theta(S_{\text{mfc}})$ captures (11.4) suggesting that the seed needs to change before quantization takes place.

Adjustments in gravity refine how the species interacts with its broader environment, subtly aiding the actualization of its adaptive goals.

11.3 Macrolevel Impact

Where Eqs. (11.4)–(11.8) address micro-level transformations, Eq. (11.9), *Impact on STEG-Fabric*, integrates these processes at a larger scale, viewing the STEG-Fabric as a more advanced computational substrate:

$$\text{TC}_{n+1} = \text{TC}_n\left(\Pi_{z\in\{S,T,E,G\}} Q_Z^{\alpha}.\mathcal{M}\left(\vec{X}, \left\{\vec{Y_i}\right\}\right).\Theta(\mathcal{A}(M1, M2, M3))\right). \tag{11.9}$$

Here

- TC_{n+1} Iterative transformation function which represents the transformed state of the STEG-Fabric after the $n+1$-th iteration to capture the cumulative impact of repeated quantizations over iterations.
- TC_n: State of the STEG-Fabric at the n-th iteration.
- $\Pi_{z\in\{s,T,E,G\}} Q_Z^{\alpha}$: Quantization contribution where each Q_z represents the quantization function of one of the four dimensions, and α is the scaling factor for each quantization dimension, reflecting its influence on the transformation.
- $\mathcal{M}\left(\vec{X}, \left\{\vec{Y_i}\right\}\right)$: Feedback and modulation, where $\vec{X}$ and $\left\{\vec{Y_i}\right\}$ represent the influence of properties on the transformation process.
- $\Theta(\mathcal{A}(M1, M2, M3))$: Activation condition with Heaviside-like function, Θ, to ensure transformation occurs when at least one of $M1$, $M2$, $M2$ is active. $\mathcal{A}(M1, M2, M3)$ is the activation state reflecting meta-layer interactions.

Taken together, these space, time, energy, and gravity quantizations define quantum certainty, a state made possible by a direct encounter with $M1$, $M2$, or $M3$. Once initiated through an activation state, the resulting quantization and development of a new meta-function illustrate the profound integration of different layers in a unified CAS. A sufficiently profound will or desire can trigger interplay across these layers, allowing deeper meta-dynamics to redirect the typical functioning of the system. Such processes are pivotal to quantum computing architectures of the type $u(W(L))$ up to $W(L)$.

11.4 Further Reading

- "Adaptation in Natural and Artificial Systems" by John H. Holland: A foundational text on genetic algorithms and their role in complex adaptive systems.
- "Hidden Order: How Adaptation Builds Complexity" by John H. Holland: A more accessible exploration of complex adaptive systems.
- "Self-Organization in Biological Systems" by Scott Camazine et al.: Explores self-organization in various biological contexts, which can provide insights into how complex structures and behaviors can emerge without central control.
- "The Selfish Gene" by Richard Dawkins: A classic book on evolutionary biology that explains how genes are the units of selection and how adaptation occurs through natural selection.
- "Evolutionary Dynamics: Exploring the Equations of Life" by Martin A. Nowak: A more mathematical treatment of evolutionary dynamics.
- "Feedback Systems: An Introduction for Scientists and Engineers" by Karl Johan Åström and Richard M. Murray: A comprehensive introduction to feedback control systems, covering both classical and modern control theory.

References

1. Malik P (2025) 20-book cosmology of light series. Amazon Author Page. Accessed 14 Jan 2025. https://www.amazon.com/author/pravirmalik
2. Malik P (2020) A light-based quantum-computational model of genetics. In: 2020 IEEE international IOT, electronics and mechatronics conference (IEMTRONICS), Vancouver, BC, Canada, pp 1–8. https://doi.org/10.1109/IEMTRONICS51293.2020.9216451
3. Malik P (2022) The role of a light-based quantum computational model in the creation of an oscillating universe. In: 2022 IEEE 12th annual computing and communication workshop and conference (CCWC), Las Vegas, NV, USA, pp 0953–0959. https://doi.org/10.1109/CCWC54503.2022.9720904
4. Malik P (2017) Quantum certainty: a mathematics of natural and sustainable human history and cosmology. Possibilities Publishing
5. Emerging Technology from the arXiv (2019) A quantum experiment suggests there's no such thing as objective reality. MIT Technology Review, 12 Mar 2019. https://www.technologyreview.com/2019/03/12/136684/a-quantum-experiment-suggests-theres-no-such-thing-as-objective-reality/
6. Emerging Technology from the arXiv. Quantum entanglement can be a measure of free will. MIT Technology Review
7. Tscharntke T, Steffan-Dewenter I, Kruess A, Thies C (2002) Characteristics of insect populations on habitat fragments: a mini review. Ecol Res 17(2):229–239
8. Arena P, Patané L, Termini PS (2010) An insect brain computational model inspired by Drosophila Melanogaster: simulation results. In: The 2010 international joint conference on neural networks (IJCNN), IEEE

Chapter 12
Quaternary Interpretation of Quantum Dynamics (QIQD)

Abstract The Quaternary Interpretation of Quantum Dynamics (QIQD) proposes a novel framework for understanding quantum phenomena, emphasizing the interconnectedness of four fundamental properties: Presence, Power, Knowledge, and Harmony. Rooted in mathematical spaces generated by imagining varying light speeds, QIQD offers a unified perspective spanning quantum particles to life itself, revealing the cosmos as a cohesive system. QIQD reinterprets core quantum concepts: superposition as the interplay of potentialities across mathematical spaces, entanglement as the deep interrelationship of all layers of reality, and uncertainty as an incomplete projection of deeper deterministic spaces onto the material plane. Central to QIQD are "seeds," foundational computational objects encoding information that gives rise to all structures. The interplay of seeds across layers, with deterministic origins and emergent material expressions, forms the computational fabric of reality. QIQD further connects the differentiation of light's properties to the emergence of space, time, energy, and gravity (STEG), each reflecting one of the four foundational properties. This fourfold structure is mirrored at subatomic, atomic, and cellular levels, demonstrating its universality. Grounded in mathematical constructs like Unified Hilbert Space, Unified Quantum Object (UQO), a modified Schrödinger equation, a generalized uncertainty relation, and the concept of "quantum certainty" achieved through feedback loops and STEG quantization, QIQD offers a new paradigm for advanced quantum computing, understanding biological systems and genetics, exploring emergent intelligence, and redefining computation as a fundamental cosmic process. Key tenets of QIQD include fourfold ontological dynamics across layers, STEG as a materialized expression of these four properties, Quantum Object Code governing quantum states, and emergent computation driven by iterative fourfold dynamics.

The Quaternary Interpretation of Quantum Dynamics (QIQD) reframes quantum dynamics as an integrative process, emphasizing the universality of four foundational

P. Malik, *Pioneering New Avenues in Quantum Technology*, Studies in Smart Technologies, https://doi.org/10.1007/978-981-96-5463-5_12

properties—Presence, Power, Knowledge, and Harmony[1].[1] Rooted in the mathematical spaces generated by imagining light traveling at different speeds, QIQD offers a cohesive framework for understanding the interconnected patterns that span quantum particles, atoms, molecules, and life. This perspective reveals the cosmos as a unified system, where microcosmic behaviors resonate with macrocosmic patterns.

QIQD offers profound reinterpretations of fundamental quantum concepts, viewing them through the lens of its fourfold framework:

- Superposition: Traditionally seen as the coexistence of multiple states, superposition within QIQD reflects the interplay of potentialities across mathematical spaces. Each space contributes its distinct character, shaping the emergent whole.
- Entanglement: In QIQD, entanglement transcends its usual definition as a connection between particles. It represents the deep interrelationship of all layers—quantum, physical, chemical, biological—of reality, where changes in one layer resonate throughout the system. Note that a deeper dive into these layers will be taken up in Chaps. 13 through 16.
- Uncertainty: QIQD reframes uncertainty not as an intrinsic limitation but as a reflection of the incomplete projection of deeper deterministic mathematical spaces onto the material plane. This view emphasizes the richness of the antecedent layers that must exist within or behind the quantum veil.

12.1 The Fourfold Framework

At the heart of QIQD lies a fourfold framework derived from mathematical spaces generated by imagining light traveling at varying speeds. These spaces represent distinct layers of reality, each contributing to the emergence of quantum and material phenomena.

- Mathematical space for infinite-speed light: This space reflects pure abstraction and symmetry, where universal properties such as presence, power, knowledge, and harmony coexist as unified entities.
- Mathematical space for intermediate-speed light: This space introduces differentiation, where structured sets of quantum properties emerge while retaining connections to the higher abstraction.
- Seed layer: Seeds arise within this space as unique combinations of quantum properties, encapsulating functions or properties that act as blueprints for materialization.
- Material space: At this level, seeds manifest as space, time, energy, and gravity, forming the structures of matter and life.

[1] The idea of QIQD is introduced in the forthcoming article published by Springer [1] and here is elaborated as per the thesis in the book.

12.2 Seeds as the Primary Object of Computation

In QIQD, seeds are the foundational computational objects. These seeds encapsulate specific functions or properties that determine their behavior and interactions across layers. By drawing from the mathematical spaces generated in the higher layers, seeds serve as the carriers of potentiality, encoding the information that gives rise to all structures in matter and life.

The interplay of seeds across layers is central to QIQD. While their origins are deeply deterministic, their expression in the material layer involves emergent dynamics, allowing for adaptability and complexity [2]. This duality of seeds—as both deterministic and emergent entities—highlights their pivotal role in the computational fabric of reality.

QIQD interprets the differentiation of light's properties as it slows from infinite speed to finite speeds, creating distinct mathematical spaces. These spaces give rise to tangible phenomena like space, time, energy, and gravity:

- Space: Emerges from the granularity of light's knowledge, forming spatial structures.
- Time: Reflects the realization of intent within seeds, overcoming opposition through power.
- Energy: Derives from the transformation of presence into dynamic forms.
- Gravity: Represents the relational harmony between seeds, shaping coherent systems.

Further symmetrical faces of the underlying fourfoldness than organize the layers of matter and life.

Subatomic dynamics at the quantum particle level mirrors the fourfold properties:

- Presence: Higgs bosons provide mass, grounding existence.
- Power: Leptons exhibit interactional potency.
- Knowledge: Quarks form the foundation of matter.
- Harmony: Gauge bosons mediate forces, ensuring coherence.

Atomic structure in Periodic Table reflects these properties:

- Presence: D-shell atoms exhibit stability and coherence.
- Power: S-shell atoms demonstrate energy-dense behaviors.
- Knowledge: P-shell atoms, archetypical of different groups, align with knowledge.
- Harmony: F-shell atoms exhibit relational qualities, fostering connections.

At the cellular level, molecular plans categorize key biological elements:

- Presence: Proteins drive cellular function, embodying presence through activity.
- Power: Polysaccharides store energy, enabling cells to perform work.
- Knowledge: Nucleic acids encode genetic information, representing knowledge.
- Harmony: Lipids create cellular compartments, promoting coherence and specialization.

12.3 The Mathematics Underpinning QIQD

The philosophical depth of QIQD is grounded in mathematical constructs that provide a rigorous framework for its layered interpretation of reality. These include the following concepts:

- Unified Hilbert Space: QIQD employs the idea of a universal Hilbert space to integrate the four quantum layers. Each layer corresponds to a distinct Hilbert space—ranging from the infinite abstraction of light at infinite speed to the concrete manifestations of space, time, energy, and gravity. The mathematical combination of these spaces using tensor products ensures a seamless representation of quantum systems.
- Unified Quantum Object (UQO): The UQO represents quantum states as they traverse the layers, encapsulating both deterministic structures and probabilistic behaviors. It serves as a bridge between the wholeness of higher layers and the fragmented realities of the material layer.
- Modified Schrödinger Equation: Within QIQD, the Schrödinger equation is reimagined to incorporate the influence of meta-layers. This reinterpretation highlights the dynamic interaction between quantum states and their layered contexts, offering insights into how form evolves from abstract potential to material manifestation.
- Generalized Uncertainty Relation: The uncertainty principle is expanded to reflect the layered dynamics of QIQD. Rather than being an intrinsic limitation, uncertainty is viewed as a projection of deeper deterministic structures onto the observable layer. This perspective aligns with the emergent properties of seeds and their interactions across layers.
- Quantum certainty: It provides a deterministic foundation within its layered framework, complementing the probabilistic nature of quantum mechanics. It arises from feedback loops connecting meta-layers ($M1$, $M2$, $M3$) with the base layer (B), enabling precise activation of meta-functions encoded in higher quantum states. Through STEG quantization, these interactions drive materialization, ensuring coordinated transformations across layers. The Unified Quantum Object (UQO) encapsulates these dynamics, integrating deterministic and probabilistic elements to bridge abstract potentialities with tangible outcomes. Quantum certainty underpins adaptive processes, allowing intentional changes at both micro- and macroscales, redefining the interplay of possibility and materialization as a mathematically grounded principle within QIQD.

12.4 QIQD Quantum Computational Architectures

QIQD reimagines quantum computing through a progressive exploration of various architectural approaches, reflecting the transition from traditional to more nuanced and advanced methodologies. Each architecture represents a step toward a deeper interaction with the quantum whole through differentiated layers, functions, and properties.

The traditional approach to quantum computing predominantly derives meaning from quantum phenomena by utilizing probability and statistics within the context of the algorithms at play. This method is primarily concerned with what can be described as the r-properties of quantum systems, deconstructing and reconstructing $W(L)$ into $r(W(L))$ as discussed in Chap. 1. In contrast, QIQD introduces enhanced quantum architectures that aim to capture more of $W(L)$'s essence, potentially evolving into $u(W(L))$ and beyond. These architectures focus on the l-properties of quantum systems, employing a different measurement philosophy that allows the emergence of a functional form referred to as $e(l)$. The envisioned architecture would fundamentally differ by focusing on "functions" rather than the properties of quantum objects within the r-set and will be further elaborated in Chaps. 17–20.

The most advanced form of quantum computational architecture envisioned to exist leveraging QIQD embraces the complexity of distinct wholes interacting with one another. This form considers the dynamics between different quantum wholes—incorporating elements such as human intent or specific emotional states—which potentially affects the quantum landscape. Complex adaptive quantum architectures require sophisticated models that are sensitive to feedback, intent, and other influencing factors. This approach integrates aspects of complex systems theory, including nano-devices technologies and adaptive systems, to create a combined quantum state represented as $(W + W_o)$ $(L + L_o)$. The malleability of 'reality' under these specific influences is highlighted, paving the way for a new paradigm in quantum computing where quantum computation incorporates human interaction as an integral component. Note that this book will discuss architectures and algorithms that include concepts to do with nano-devices and adaptive systems as discussed in Chaps. 20–24, but will not specifically address human interaction and intent. The notion of 'intent' was covered in Chap. 11 and would need to be further extrapolated given the materials in Chaps. 20–24 to arrive at human intent based $(W + W_o)$ $(L + L_o)$ quantum computers.

12.5 Applications and Future Directions

The conceptual and computational power of seeds within the QIQD framework paves the way for transformative applications across various domains. A non-technical series of articles exploring this is also available on Forbes [3]:

- Advanced Quantum Computing: Seeds provide a new computational paradigm, enabling multi-layered quantum systems that leverage deterministic blueprints alongside emergent probabilistic behaviors. This will revolutionize quantum algorithms and hardware design.

- Biological Systems and Genetics: The seed concept aligns closely with the mechanisms of genetic encoding and adaptation. QIQD will inform new models for understanding genetic evolution and bioengineering [4], potentially unlocking methods to manipulate genetic properties at quantum levels [5].
- Emergent Intelligence: Seeds, as computational entities, provide a theoretical foundation for exploring the origins of intelligence and consciousness [6, 7]. By mapping the progression of seed functions across layers, QIQD offers insights into the emergence of cognitive and adaptive systems [8].
- Philosophical Extensions: Beyond practical applications, QIQD redefines the nature of computation itself. It suggests that computation is not merely a mechanical process but a fundamental characteristic of the cosmos [9], manifesting through the properties encoded in seeds [10].

12.6 Further Reading

- The Fractal Geometry of Nature by Benoit Mandelbrot: A classic text by the "father of fractals," this book explores the ubiquitous nature of fractals in the natural world and their mathematical properties.
- Quantum Effects in Biology edited by Masoud Mohseni, Yasser Omar, Gregory S. Engel, and Martin B. Plenio: This more technical book delves deeper into the specific quantum mechanisms proposed in various biological systems. It covers topics like quantum coherence in photosynthesis, quantum effects in enzyme catalysis, and the role of quantum mechanics in DNA mutations. It's suitable for readers with some background in physics and biology.
- The Rainbow and the Worm: The Physics of Organisms by Mae-Wan Ho: This book takes a broader perspective, exploring the interplay between quantum physics, biology, and the environment. It discusses how organisms might utilize quantum coherence and entanglement to achieve remarkable efficiency and coordination in their functions.
- "What Is Computation?" edited by John Haugeland: A collection of essays exploring different philosophical perspectives on the nature of computation.
- "The Computational Theory of Mind" by Stephen Stich: Explores the idea that the mind is a computational system.

References

1. Malik P (2025) An imaginative inquiry into a quaternary interpretation of quantum dynamics & its technological implications. In: Bradford PG, Gadsden SA, Koul SK, Ghatak KP (eds) Proceedings of IEMTRONICS 2024: international IoT, electronics and mechatronics conference, volume 1. Lecture notes in electrical engineering, vol 1229. Springer, Singapore
2. 't Hooft G (2020) Deterministic quantum mechanics: the mathematical equations. Front Phys 8:253 (2020)

3. Malik P (2025) Forbes councils. Accessed 23 Jan 2025. https://councils.forbes.com/profile/Pravir-Malik-Founder-QIQuantum/2258eb58-6350-495f-8d26-16581994fcc4
4. Davies P (2004) Emergent biological principles and the computational cosmos. Complexity 10(2):11–15
5. Narayanan A (1997) Quantum computing for modeling genetic networks. Biosystems 44(1):43–48
6. Tononi G (2004) An information integration theory of consciousness. BMC Neurosci 5(1):42
7. Penrose R (1994) Shadows of the mind: a search for the missing science of consciousness. Oxford University Press
8. Varela FJ, Maturana HR, Uribe R (1974) Autopoiesis: the organization of living systems, its characterization and a model. Biosystems 5(4):187–196
9. Lloyd S (2006) Programming the universe: a quantum computer scientist takes on the cosmos. Knopf
10. Wheeler JA (1990) Information, physics, quantum: the search for links. In: Zurek WH (ed) Complexity, entropy and the physics of information. Addison-Wesley, pp 3–28

Part IV
QIQD Dynamics at Different Levels of Granularity

Chapter 13
QIQD Dynamics at the Level of the Electromagnetic Spectrum

Abstract This chapter explores the emergence of structure within the electromagnetic spectrum (EMS) through the lens of the Quaternary Interpretation of Quantum Dynamics (QIQD). QIQD posits four fundamental properties—Knowledge, Presence, Harmony, and Power—that manifest dynamically within the EMS as wavelength, mass-potential, the speed of light (c), and frequency, respectively. The EMS's vast range of wavelengths represents a breadth of cosmic-level knowledge, while the constancy of c provides a foundation for harmony within which the variation of space and time can take place. Frequency, directly related to energy through Planck's equation, reflects the power aspect. The presence aspect is related to the mass-potential resident in the EMS. A proposed equation captures the cumulative contribution of "seeds" (y) to EMS properties (x) through layered dynamics, revealing the deterministic organization within the spectrum. This fourfold structure is further exemplified within the visible light spectrum, where red, yellow, green, and blue are associated with Presence, Power, Harmony, and Knowledge, respectively, demonstrating a fractal connection between color and these fundamental properties.

At the level of the electromagnetic spectrum (EMS), structure emerges through the coordinated interplay of four fundamental aspects of EMS: wavelength, mass-potential, speed of light (c), and frequency. Within the QIQD framework, these aspects of EMS reflect the dynamic manifestation of Knowledge, Presence, Harmony, and Power, respectively, core universal properties at the quantum scale.[1] Here, each EMS aspect acts as an embodiment of one of these properties and is represented by Eq. (13.1) *EMS Aspect as m-Operator Output*:

$$\begin{aligned}\text{EMS Aspect} &= m(X_{\text{ems}}) = m(\text{Pr}, \text{Po}, K, H) \\ &\rightarrow \{\text{Wavelength, Mass} - \text{Potential}, c, \text{Frequency}\}\end{aligned} \tag{13.1}$$

[1] Initial renderings of this idea appeared across the series of Cosmology of Light books referred to in [95] and then subsequently in several IEEE papers—https://ieeexplore.ieee.org/author/37086022058, and here has been further developed as per the overarching quantum-based mathematical framework.

P. Malik, *Pioneering New Avenues in Quantum Technology*, Studies in Smart Technologies, https://doi.org/10.1007/978-981-96-5463-5_13

Here '*m*' refers to the property generator from $M3$-dynamics. X refers to the archetype of information-energy that embodies the essence of organization as introduced in Chap. 4, Mathematics of Qualified Determinism, in association with UQO(X) (4.19).

The following sections surface QIQD equations for each of the quantum objects. Finally, each of the quantum objects is unified in a single QIQD representation.

13.1 Wavelength

The electromagnetic (EM) spectrum embodies an immense breadth of knowledge, encompassing a wide array of phenomena across the cosmos [1]. Spanning from radio waves to gamma rays [2], each type of wave within the spectrum is distinguished by its unique frequency and wavelength. Together, these diverse waves form a set of fundamental archetypes, reflecting the underlying patterns and principles that govern the universe.

Equations (13.2)–(13.4) highlight the wavelength-related *m*-, *n*-, and *o*-operators initially depicted in UQO (4.5).

Hence, Eq. (13.2), *Wavelength* (λ) *Correlation with Knowledge*:

$$m\left(X_{\text{ems}}^{(K)}\right) = \lambda \propto f(\text{Knowledge}). \tag{13.2}$$

Equation (13.3), *Wavelength* (λ) *Set as n-Operator Output*:

$$n\left(X_{\text{ems}}^{(K)}\right) = \lambda_{\text{set}} \ni \{\text{Archetype, Identity, Potentiality} \ldots\}. \tag{13.3}$$

Equation (13.4), *Wavelength* (λ) *Configuration as o-Operator Output*, is the result of the *o*-operator acting at $M1$:

$$\lambda_{\text{seed}} = o\left(X_x^{(K)}\right) = \sum_{x=1}^{N} h\left(a_x X_x^{(K)} + \sum_{y=1}^{M} b_y Y_{y,x}^{(K,\text{Pr},\text{Po},H)}\right). \tag{13.4}$$

Here

- $o\left(X_x^{(K)}\right)$: *o*-operator output acting on primary Knowledge elements $X_x^{(K)}$.
- $Y_{y,x}^{(K,\text{Pr},\text{Po},H)}$: Secondary contributions Knowledge, Presence, Power, and Harmony.
- $a_x > b_y$: Weighting factors for primary and secondary components.

13.2 Mass-Potential

Equations (13.5)–(13.7) highlight the mass-potential $\left(\frac{h\nu}{c^2}\right)$ m-, n-, and o-operators.

The energy-mass relationship [3], expressed through Einstein's famous equation $E = mc^2$, and the equivalence of energy (E) with $h\nu$, highlights how EMS embodies Presence. By combining these two foundational principles, we derive Eq. (13.5), *Mass-Potential* $\left(\frac{h\nu}{c^2}\right)$ *Correlation with Presence*:

$$m\left(X_{\text{ems}}^{(\text{Pr})}\right) = \frac{h\nu}{c^2} \propto f(\text{Presence}). \tag{13.5}$$

This relationship links the energy of electromagnetic waves ($h\nu$), their mass-potential (m), and the speed of light (c) to the concept of Presence. It reflects how EMS anchors substantiality, enabling the manifestation of mass through energy.

Equation (13.6), *Mass-Potential* $\left(\frac{h\nu}{c^2}\right)$ *Set as n-Operator Output*:

$$n\left(X_{\text{ems}}^{(\text{Pr})}\right) = \frac{h\nu}{c^2}_{\text{set}} \ni \{\text{Mass energy equivalence, Substantiality, } \ldots\}. \tag{13.6}$$

Equation (13.7), *Mass-Potential* $\left(\frac{h\nu}{c^2}\right)$ *Seed Configuration as o-Operator Output*, is the result of the o-operator acting at $M1$:

$$\frac{h\nu}{c^2}_{\text{seed}} = o\left(X_x^{(\text{Pr})}\right) = \sum_{x=1}^{N} h\left(a_x X_x^{(\text{Pr})} + \sum_{y=1}^{M} b_y Y_{y,x}^{(K,\text{Pr},\text{Po},H)}\right). \tag{13.7}$$

Here

- $o\left(X_x^{(\text{Pr})}\right)$: o-operator output acting on primary Presence elements $X_x^{(\text{Pr})}$.
- $Y_{y,x}^{(K,\text{Pr},\text{Po},H)}$: Secondary contributions Knowledge, Presence, Power, and Harmony.
- $a_x > b_y$: Weighting factors for primary and secondary components.

13.3 Speed of Light (c)

Equations (13.8)–(13.10) highlight the speed of light (c) m-, n-, and o-operators.

The constancy of the speed of light (c) serves as a cornerstone for the principle of Harmony within EMS. The speed of light remains invariant in a vacuum, providing a stable framework that allows for the variation of space and time. This constancy enables a harmonious interplay of forms and relationships within the universe. The embodiment of Harmony in EMS is captured by Eq. (13.8), *c Correlation with Harmony*, follows:

$$m\left(X_{\text{ems}}^{(H)}\right) = c \propto f(\text{Harmony}). \tag{13.8}$$

Here, c, as the universal constant of light speed, represents the stabilizing force that allows for coherent relationships and interactions within EMS.

Equation (13.9), *c Set as n-Operator Output*:

$$n\left(X_{\text{ems}}^{(H)}\right) = c_{\text{set}} \ni \{\text{Stabilization, Relational Coherence, Temporal} \\ - \text{Spatial Unity} \ldots\}. \tag{13.9}$$

Equation (13.10), *c Seed Configuration as o-Operator Output*, is the result of the o-operator acting at $M1$:

$$c_{\text{seed}} = o\left(X_x^{\{H\}}\right) = \sum_{x=1}^{N} h\left(a_x X_x^{\{H\}} + \sum_{y=1}^{M} b_y Y_{y,x}^{\{K,\text{Pr},\text{Po},H\}}\right). \tag{13.10}$$

Here

- $o\left(X_x^{\{H\}}\right)$: o-operator output acting on primary Harmony elements $X_x^{\{H\}}$.
- $Y_{y,x}^{\{K,\text{Pr},\text{Po},H\}}$: Secondary contributions Knowledge, Presence, Power, and Harmony.
- $a_x > b_y$: Weighting factors for primary and secondary components.

13.4 Frequency

Equations (13.11)–(13.13) highlight the frequency m-, n-, and o-operators.

EMS exhibits a wide range of frequencies, which directly correlate with the energy levels of electromagnetic waves [4]. Higher frequencies, such as those of gamma rays, correspond to greater energy, while lower frequencies, such as radio waves, correspond to lower energy. The relationship between frequency (ν) and energy is defined by Planck's equation, where h is Planck's constant. This relationship is embodied by Power within EMS as depicted by Eq. (13.11), *hυ Correlation with Power*, follows:

$$m\left(X_{\text{ems}}^{(\text{Po})}\right) = h\upsilon \propto f(\text{Power}). \tag{13.11}$$

This equation demonstrates how the varying frequencies of EMS encode the principle of Power, highlighting the dynamic flow of energy across the spectrum.

Equation (13.12), *hυ Set as n-Operator Output*:

$$n\left(X_{\text{ems}}^{(\text{Po})}\right) = h\upsilon_{\text{set}} \ni \{\text{Dynamic energy flow, Power variability}, \ldots\}. \tag{13.12}$$

Equation (13.13), *hυ Seed Configuration as o-Operator Output*, is the result of the o-operator acting at $M1$:

$$h\upsilon_{\text{seed}} = o\left(X_x^{(\text{Po})}\right) = \sum_{x=1}^{N} h\left(a_x X_x^{(\text{Po})} + \sum_{y=1}^{M} b_y Y_{y,x}^{(K,\text{Pr},\text{Po},H)}\right). \tag{13.13}$$

Here

- $o\left(X_x^{(\text{Po})}\right)$: o-operator output acting on primary Harmony elements $X_x^{(\text{Po})}$.
- $Y_{y,x}^{(K,\text{Pr},\text{Po},H)}$: Secondary contributions Knowledge, Presence, Power, and Harmony.
- $a_x > b_y$: Weighting factors for primary and secondary components.

13.5 Unified QIQD Representation

A unified QIQD representation at the level of electromagnetic spectrum is depicted by Eq. (13.14), Electromagnetic Spectrum Unified QIQD:

$$Q_{\text{ems}} = \sum_{x\in\{K,\text{Pr},\text{Po},H\}} \sum_{y\in\{1,\ldots,N\}} c_{x,y}\left(o\left(X_y^{(x)}\right) \wedge n\left(X_y^{(x)}\right) \wedge m\left(X_y^{(x)}\right)\right). \tag{13.14}$$

Here

- $X_y^{(x)}$: Represents the primary components of Knowledge, Presence, Power, and Harmony.
- o, n, m: Operators corresponding to $M1$ (seed level), $M2$ (pure property level), and $M3$ (complex agglomerate generation level) dynamics.
- $c_{x,y}$: Coefficients reflecting amplitude contributions of each property.

The equation reflects the cumulative contributions of all seeds (y) to the properties (x) of the electromagnetic spectrum. By tracing each $X_y^{(x)}$ through the layered dynamics of the o-, n-, and m-operators, the equation captures the deterministic organization encoded in the electromagnetic spectrum.

Note that studies on the relationship between color and property reveal a similar fractal connection, where the four organizing energies manifest within visible light—a small subset of the electromagnetic spectrum's vast wavelength range [5]. Specifically, 'red,' corresponding to wavelengths from 635 to 760 nm, aligns with the concept of 'presence.' 'Yellow,' associated with wavelengths from 570 to 590 nm, resonates with the concept of 'power.' Similarly, 'green,' spanning wavelengths from 520 to 550 nm, is linked to the concept of 'harmony,' while 'blue,' covering wavelengths from 460 to 490 nm, corresponds to the concept of 'knowledge'.

Equations that mirror (13.2)–(13.4) for blue, (13.5)–(13.7) for red, (13.8)–(13.10) for green, and (13.11)–(13.13) for yellow can similarly be derived. Equation (13.14) can be made to just focus on colors by restricting the range of what K, H, Pr, Po depict.

Hence

- x represents the property associated with each color:

(a) *K* (Knowledge): Blue (460–490 nm).
(b) *H* (Harmony): Green (520–550 nm).
(c) Po (Power): Yellow (570–590 nm).
(d) Pr (Presence): Red (635–760 nm).

13.6 Further Reading

- "Cosmology" by Steven Weinberg: This book provides a detailed overview of cosmology, including discussions of how different wavelengths of the EMS reveal information about the universe.
- "Optics" by Eugene Hecht: A comprehensive textbook on optics, including a discussion of the physics of color.
- "Interaction of Color" by Josef Albers: A classic book on color theory from an artistic perspective.

References

1. Griffiths DJ (2017) Introduction to electrodynamics, 4th edn. Cambridge University Press
2. NASA (2025) The electromagnetic spectrum. In: NASA science: share the science. Accessed 17 Jan 2025. https://science.nasa.gov/ems/
3. Einstein A (1905) Does the inertia of a body depend upon its energy content? Ann Phys 18:639–641
4. The electromagnetic spectrum. In: CHM101: chemistry and global awareness. Furman University. Accessed 17 Jan 2025. https://chem.libretexts.org/Courses/Furman_University/CHM101:_Chemistry_and_Global_Awareness_(Gordon)/05:_Basics_of_Nuclear_Science/5.02:_The_Electromagnetic_Spectrum
5. Malacara D (2011) Color vision and colorimetry: theory and applications, 2nd edn. SPIE Press, Bellingham, WA

Chapter 14
QIQD Dynamics at the Level of Quantum Particles

Abstract This chapter leverages Quaternary Interpretation of Quantum Dynamics (QIQD), to categorize and unify fundamental particles within the Standard Model. QIQD proposes that the four known particle groups—quarks, leptons, bosons, and the Higgs boson—represent dynamic manifestations of four fundamental properties: Knowledge, Power, Harmony, and Presence, respectively. This relationship is formalized through a primary equation, which expresses particle categories as outputs of a property generator (m) acting on an informational-energetic archetype (X). The chapter further posits that quarks that make up protons, which in turn determine atomic number and thus elemental identity, embody Knowledge. The Higgs boson, responsible for imparting mass to quarks through interaction with the Higgs field, represents Presence. Bosons, mediating fundamental forces (strong, electromagnetic, and weak), are associated with Harmony. Finally, leptons, exemplified by the electron and their role in energy transfer, embody Power. The chapter derives specific QIQD equations for each particle type, demonstrating how these properties manifest within their respective characteristics. Furthermore, it explores the internal fourfold structure within the boson category itself, assigning Knowledge to gluons, Presence to photons, Harmony to the hypothetical graviton, and Power to W and Z bosons. This unified QIQD representation offers a new perspective on the organization of fundamental particles and their interrelationships.

All known quantum particles are organized into one of four groups—quarks, leptons, bosons, and Higgs boson—as per the Standard Model [1]. Within the QIQD framework, these reflect the dynamic manifestation of Knowledge, Power, Harmony, and Presence respectively[1] as summarized by Eq. (14.1) *Quantum Particle Categories as m-Operator Output*:

$$\text{Quantum Particle Categories} = m\left(X_{qp}\right) = m(\text{Pr}, \text{Po}, K, H)$$

[1] Initial renderings of this idea appeared across the series of Cosmology of Light books referred to in [95] and then subsequently in several IEEE papers—https://ieeexplore.ieee.org/author/37086022058, and here has been further developed as per the overarching quantum-based mathematical framework.

P. Malik, *Pioneering New Avenues in Quantum Technology*, Studies in Smart Technologies, https://doi.org/10.1007/978-981-96-5463-5_14

$$\rightarrow \{\text{Higgs} - \text{boson, Lepton, Quark, Boson}\}. \quad (14.1)$$

Here 'm' refers to the property generator from $M3$-dynamics. X refers to the archetype of information-energy that embodies the essence of organization as introduced in association with UQO(X) (4.19).

The following sections surface QIQD equations for each type of quantum particle. Finally, quantum particles are unified in a single QIQD representation.

14.1 Quarks

At the quantum level, quarks have been hypothesized to represent a form of knowledge. This idea arises from the observation that protons, which determine an element's atomic number, are composed of two "up" quarks and one "down" quark [2]. The atomic number uniquely identifies each element in the Periodic Table; for example, an atomic number of 47 corresponds unequivocally to silver. This connection implies that the unique properties of an element—including its identity and predictable behavior in the universe—are intrinsically tied to its quark composition.

Equations (14.2)–(14.4) highlight quark-related m-, n-, and o-operators initially depicted in the UQO (4.5).

Hence, Eq. (14.2), Quark Correlation with Knowledge:

$$m\left(X_{qp}^{(K)}\right) = \text{Quark} \propto f(\text{Knowledge}). \quad (14.2)$$

Equation (14.3), Quark Set as n-Operator Output:

$$n\left(X_{qp}^{(K)}\right) = \text{Quark}_{\text{set}} \ni \{\text{Atomic identity, subatomic stability} \ldots\}. \quad (14.3)$$

Equation (14.4), Quark Seed Configuration as o-Operator Output, is the result of the o-operator acting at $M1$:

$$\text{Quark}_{\text{seed}} = o\left(X_x^{(K)}\right) = \sum_{x=1}^{N} h\left(a_x X_x^{(K)} + \sum_{y=1}^{M} b_y Y_{y,x}^{(K,\text{Pr},\text{Po},H)}\right). \quad (14.4)$$

Here

- $o\left(X_x^{(K)}\right)$: o-operator output acting on primary Knowledge elements $X_x^{(K)}$.
- $Y_{y,x}^{\{K,\text{Pr},\text{Po},H\}}$: Secondary contributions Knowledge, Presence, Power, and Harmony.
- $a_x > b_y$: Weighting factors for primary and secondary components.

14.2 Higgs Boson

Among the fundamental particles, the Higgs boson [3] stands out as a unique entity. In ordinary matter, the mass of an atom resides primarily in its nucleus, which consists of protons and neutrons, each composed of three quarks. However, it is the interaction with the Higgs field that grants mass to these quarks. As such, the Higgs boson can be regarded as the "mass giver," the particle responsible for imparting substance to quarks.

Equations (14.5)–(14.7) highlight Higgs boson-related *m*-, *n*-, and *o*-operators. Hence, Eq. (14.5), Higgs boson Correlation with Presence:

$$m\left(X_{qp}^{(\mathrm{Pr})}\right) = \mathrm{Higgs} - \mathrm{boson} \propto f(\mathrm{Presence}) \tag{14.5}$$

Equation (14.6), Higgs boson Set as *n*-Operator Output:

$$n\left(X_{qp}^{(Pr)}\right) = \mathrm{Higgs} - \mathrm{boson}_{\mathrm{set}} \ni \{\mathrm{Mass\ generation,\ Substantiality,}\ \mathrm{Stabilizing\ matter}, \ldots\} \tag{14.6}$$

Equation (14.7), Higgs boson Seed Configuration as *o*-Operator Output, is the result of the *o*-operator acting at $M1$:

$$\mathrm{Higgs} - \mathrm{boson}_{\mathrm{seed}} = o\left(X_x^{(\mathrm{Pr})}\right) = \sum_{x=1}^{N} h\left(a_x X_x^{(\mathrm{Pr})} + \sum_{y=1}^{M} b_y Y_{y,x}^{(K,\mathrm{Pr},\mathrm{Po},H)}\right) \tag{14.7}$$

Here

- $o\left(X_x^{(\mathrm{Pr})}\right)$: *o*-operator output acting on primary Presence elements $X_x^{(\mathrm{Pr})}$.
- $Y_{y,x}^{(K,\mathrm{Pr},\mathrm{Po},H)}$: Secondary contributions Knowledge, Presence, Power, and Harmony.
- $a_x > b_y$: Weighting factors for primary and secondary components.

14.3 Boson

Bosons, the fundamental carriers of force, serve as the "glue" enabling interactions between matter particles [4]. Among the three key bosons, the gluon mediates the strong nuclear force, binding quarks within protons and neutrons. The photon governs the electromagnetic force, responsible for electrical and magnetic interactions. Meanwhile, the W and Z bosons [5, 6] mediate the weak nuclear force, which is crucial for radioactive decay processes. Bosons, therefore, embody the principles of quantum harmony, facilitating the forces that underpin the universe.

Equations (14.8)–(14.10) highlight the Boson-related *m*-, *n*-, and *o*-operators. Hence, Eq. (14.8), Boson Correlation with Harmony, follows:

$$m\left(X_{qp}^{(H)}\right) = \text{Boson} \propto f(\text{Harmony}). \tag{14.8}$$

Equation (14.9), Boson Set as *n*-Operator Output:

$$n\left(X_{qp}^{(H)}\right) = \text{Boson}_{\text{set}} \ni \{\text{Force mediation, Strong nuclear bond, Electromagnetic interaction}, \ldots\}. \tag{14.9}$$

Equation (14.10), Boson Seed Configuration as *o*-Operator Output, is the result of the *o*-operator acting at $M1$:

$$\text{Boson}_{\text{seed}} = o\left(X_x^{\{H\}}\right) = \sum_{x=1}^{N} h\left(a_x X_x^{\{H\}} + \sum_{y=1}^{M} b_y Y_{y,x}^{\{K,\text{Pr},\text{Po},H\}}\right). \tag{14.10}$$

Here

- $o\left(X_x^{\{H\}}\right)$: *o*-operator output acting on primary Harmony elements $X_x^{\{H\}}$.
- $Y_{y,x}^{\{K,\text{Pr},\text{Po},H\}}$: Secondary contributions Knowledge, Presence, Power, and Harmony.
- $a_x > b_y$: Weighting factors for primary and secondary components.

14.4 Lepton

Insight into the category of leptons arises by considering the electron as a surrogate of this category. Electrons are fundamental particles that exhibit the characteristic of energy. Unlike quarks, which only exist within composite particles, leptons exist as individual entities. Leptons are considered point-like particles, lacking any discernible internal structure [7]. The electron is the most well-known lepton. Arabatzis [8] delves into the properties of electrons in his work, "Representing Electrons: A Biographical Approach to Theoretical Entities." The electron serves as a representative of the lepton class. Notably, the electron is linked to the transfer of energy and power, along with other properties.

Equations (14.11)–(14.13) highlight the Lepton-related *m*-, *n*-, and *o*-operators.

Hence, Eq. (14.11), Lepton Correlation with Power, follows:

$$m\left(X_{qp}^{(\text{Po})}\right) = \text{Lepton} \propto f(\text{Power}). \tag{14.11}$$

Equation (14.12), Lepton Set as *n*-Operator Output:

$$n\left(X_{qp}^{(\text{Po})}\right) = \text{Lepton}_{\text{set}} \ni \{\text{Energy transfer, Stability, Point} - \text{like existence}, \ldots\}. \tag{14.12}$$

Equation (14.13), Lepton Seed Configuration as *o*-Operator Output, is the result of the *o*-operator acting at $M1$:

$$\text{Lepton}_{\text{seed}} = o\left(X_x^{(\text{Po})}\right) = \sum_{x=1}^{N} h\left(a_x X_x^{(\text{Po})} + \sum_{y=1}^{M} b_y Y_{y,x}^{(K,\text{Pr},\text{Po},H)}\right). \tag{14.13}$$

Here

- $o\left(X_x^{(\text{Po})}\right)$: *o*-operator output acting on primary Harmony elements $X_x^{(\text{Po})}$.
- $Y_{y,x}^{(K,\text{Pr},\text{Po},H)}$: Secondary contributions Knowledge, Presence, Power, and Harmony.
- $a_x > b_y$: Weighting factors for primary and secondary components.

14.5 Unified QIQD Representation

A unified QIQD representation at the level of quantum particles is depicted by Eq. (14.14), Quantum Particle Unified QIQD:

$$Q_{qp} = \sum_{x\in\{K,\text{Pr},\text{Po},H\}} \sum_{y\in\{1,\ldots,N\}} c_{x,y}\left(o\left(X_y^{(x)}\right) \wedge n\left(X_y^{(x)}\right) \wedge m\left(X_y^{(x)}\right)\right). \tag{14.14}$$

Here

- $X_y^{(x)}$: Represents the primary components of Knowledge, Presence, Power, and Harmony.
- *o*, *n*, *m*: Operators corresponding to $M1$ (seed level), $M2$ (pure property level), and $M3$ (complex agglomerate generation level) dynamics.
- $c_{x,y}$: Coefficients reflecting amplitude contributions of each property.

The equation reflects the cumulative contributions of all seeds (y) to the properties (x) of quantum particles. By tracing each $X_y^{(x)}$ through the layered dynamics of the *o*-, *n*-, and *m*-operators, the equation captures the deterministic organization encoded at the quantum particle level.

Note that the Boson category can also be perceived to house the fourfold structure. Equations that mirror (14.2)–(14.4), i.e., Knowledge) for gluons, (14.5)–(14.7), i.e., Presence) for photons, (14.8)–(14.10), i.e., Harmony) for the hypothetical graviton, and [(14.11)–(14.13), i.e., Power] for W and Z bosons, can similarly be derived. Equation (14.14) can also be made to just focus on bosons by restricting the range of what K, H, Pr, Po depict.

Hence,

- x represents the property associated with each boson:

  (a) K (Knowledge): Gluon.
  (b) H (Harmony): Graviton.
  (c) Po (Power): W&Z Boson.

(d) Pr (Presence): Photon.

14.6 Further Reading

- "Quarks and Leptons: An Introductory Course in Modern Particle Physics" by Halzen, F., and A. D. Martin: A standard text for understanding how quarks combine to form protons, neutrons, and other hadrons.
- "The Periodic Table: Its Story and Its Significance" by Scerri: Discusses how the atomic number (i.e., the number of protons) defines an element's place in the periodic table and thus its identity.
- "The Tao of Physics: An Exploration of the Parallels Between Modern Physics and Eastern Mysticism" by Capra, F: While not a primary physics text, it discusses broader philosophical views on quantum theory and the idea that fundamental particles (including quarks) might carry "information" or "knowledge" about reality.
- "The Higgs Hunter's Guide" by Gunion, J. F., H. E. Haber, G. L. Kane, and S. Dawson: A comprehensive resource on Higgs physics, detailing how the Higgs field interacts with fermions (including quarks) and imparts mass.
- "An Introduction to Quantum Field Theory" by Peskin, M. E., and D. V. Schroeder: Classic quantum field theory text that explains the mathematical framework behind the Standard Model, including quark mass generation via Yukawa coupling to the Higgs field.
- "Gauge Theories in Particle Physics" by I. J. R. Aitchison and A. J. G. Hey: A comprehensive text on gauge theories, which are the foundation of the Standard Model's description of forces.
- "Decoding Reality: The Universe as Quantum Information" by Vlatko Vedral: Explores the idea that the universe can be understood as fundamentally informational.

References

1. Particle Data Group (2022) Review of particle physics. Progr Theoret Exp Phys 8:083C01
2. Gell-Mann M (1964) A schematic model of baryons and mesons. Phys Lett 8(3):214–215
3. Higgs PW (1964) Broken symmetries and the masses of gauge bosons. Phys Rev Lett 13(16):508–509
4. Gaillard MK, Grannis PD, Sciulli FJ (1999) The standard model of particle physics. Rev Mod Phys 71(2):S96–S111
5. Arnison G et al (UA1 Collaboration) (1983) Experimental observation of isolated large transverse energy electrons with associated missing energy at $\sqrt{s} = 540$ GeV. Phys Lett B 122(1):103–116
6. Banner M et al (UA2 Collaboration) (1983) Observation of single isolated electrons of high transverse momentum in events with missing transverse energy at the CERN SPS collider. Phys Lett B 122(5):476–485

7. Perkins DH (2000) Introduction to high energy physics, 4th edn. Cambridge University Press, Cambridge
8. Arabatzis T (2005) Representing electrons: a biographical approach to theoretical entities. University of Chicago Press, Chicago

Chapter 15
QIQD Dynamics at the Level of Atoms

Abstract This chapter leverages Quaternary Interpretation of Quantum Dynamics (QIQD), to categorize and unify elements within the Periodic Table. QIQD proposes that the four main atomic groups—s-block, p-block, d-block, and f-block—represent dynamic manifestations of four fundamental properties: Power, Knowledge, Presence, and Harmony, respectively. It posits that the p-block, with its versatile orbital configurations and role in key elements like carbon and silicon, embodies Knowledge. The d-block, encompassing transition metals crucial for technological and structural applications, represents Presence, emphasizing adaptability and stability. The f-block, with its multi-lobed orbitals fostering interconnectedness and experiments in collective stability among lanthanides and actinides, embodies Harmony. Finally, the s-block, characterized by high reactivity and energy release in elements like hydrogen and alkali metals, reflects Power. The chapter derives specific QIQD equations for each atomic group, highlighting how these properties manifest within their respective characteristics. A unified QIQD representation at the atomic level captures the cumulative contributions of "seeds" (y) to atomic properties (x) through layered dynamics of o-, n-, and m-operators. This unified perspective offers a new understanding of the organization of elements and their inherent properties within the Periodic Table.

All known atoms are organized into one of four groups—p-Group, s-Group, d-Group, and f-Group—as per the Periodic Table [1]. Within the QIQD framework, these reflect the dynamic manifestation of Knowledge, Power, Presence, and Harmony, respectively[1] as summarized by Eq. (15.1) Atom Categories as m-Operator Output:

$$\begin{aligned}\text{Atom Categories} &= m(X_{\text{atom}}) = m(\text{Pr, Po, K, H}) \\ &\rightarrow \{\text{p-Group, s-Group, d-Group, f-Group}\}. \end{aligned} \tag{15.1}$$

[1] Initial renderings of this idea appeared across the series of Cosmology of Light books referred to earlier and then subsequently in several IEEE papers—https://ieeexplore.ieee.org/author/37086022058, and here has been further developed as per the overarching quantum-based mathematical framework.

P. Malik, *Pioneering New Avenues in Quantum Technology*, Studies in Smart Technologies, https://doi.org/10.1007/978-981-96-5463-5_15

Here '*m*' refers to the property generator from M3-dynamics. *X* refers to the archetype of information-energy that embodies the essence of organization as introduced in association with UQO(*X*) (4.19).

The following sections surface QIQD equations for each type of atom category. Finally, atoms are unified in a single QIQD representation.

15.1 p-Group

The p-Group's versatility reflects Knowledge as the archetype of elemental possibilities. The equal probability of finding an electron in either of two orbital lobes [2] creates an essential polarity that functions as a quantum-level switch. This is foundational in inviting other switches and is instrumental in the archetypal possibilities represented by this group. It also houses elements such as carbon—the foundation of DNA and life-forms [3], and silicon—a key component in virtual 'thinking' machines [4].

Equations (15.2–15.4) highlight the p-Group-related *m*-, *n*-, and *o*-operators initially depicted in the UQO (4.5).

Hence, Eq. (15.2), p-Group Correlation with Knowledge:

$$m\left(X_{\text{atom}}^{(\text{K})}\right) = \text{p-Group} \propto f(\text{Knowledge}). \tag{15.2}$$

Equation (15.3), p-Group Set as *n*-Operator Output:

$$n\left(X_{\text{atom}}^{(\text{K})}\right) = \text{p-Group}_{\text{set}} \ni \{\text{Archetypal possibilities, bases of thinking} \ldots .\} \tag{15.3}$$

Equation (15.4), p-Group Seed Configuration as o-Operator Output, is the result of the o-operator acting at M1:

$$\text{p-Group}_{\text{seed}} = o\left(X_x^{(\text{K})}\right) = \sum_{x=1}^{N} h\left(a_x X_x^{(\text{K})} + \sum_{y=1}^{M} b_y Y_{y,x}^{(\text{K, Pr,Po,H})}\right). \tag{15.4}$$

Here:

- $o\left(X_x^{(\text{K})}\right)$: o-operator output acting on primary Knowledge elements $X_x^{(\text{K})}$.
- $Y_{y,x}^{\{\text{K, Pr,Po,H}\}}$: Secondary contributions Knowledge, Presence, Power, and Harmony.
- $a_x > b_y$: Weighting factors for primary and secondary components.

15.2 d-Group

The d-Group embodies Presence, reflecting adaptability, stability, and strength. These transition metals, such as Iron Nickel, Copper, Silver, and so on, play pivotal roles in technological processes, industrial stability, and structural resilience [5].

Equations (15.5–15.7) highlight the d-Group-related *m*-, *n*-, and *o*-operators.

Hence, Eq. (15.5), d-Group Correlation with Presence:

$$m\left(X_{\text{atom}}^{(\text{Pr})}\right) = \text{d-Group} \propto f(\text{Presence}) \tag{15.5}$$

Equation (15.6), d-Group Set as *n*-Operator Output:

$$n\left(X_{\text{atom}}^{(\text{Pr})}\right) = \text{d-Group}_{\text{set}} \ni \{\text{Stability, Strength, Adaptability} \ldots.\}. \tag{15.6}$$

Equation (15.7), d-Group Seed Configuration as o-Operator Output, is the result of the o-operator acting at M1:

$$\text{d-Group}_{\text{seed}} = o\left(X_x^{(\text{Pr})}\right) = \sum_{x=1}^{N} h\left(a_x X_x^{(\text{Pr})} + \sum_{y=1}^{M} b_y Y_{y,x}^{(\text{K, Pr,Po,H})}\right). \tag{15.7}$$

Here:

- $o\left(X_x^{(\text{Pr})}\right)$: o-operator output acting on primary Presence elements $X_x^{(\text{Pr})}$.
- $Y_{y,x}^{(\text{K, Pr,Po,H})}$: Secondary contributions Knowledge, Presence, Power, and Harmony.
- $a_x > b_y$: Weighting factors for primary and secondary components.

15.3 f-Group

The f-Group embodies Harmony, with its multi-lobed orbitals creating interconnected bonds and collective stability. These elements—lanthanides [6] and actinides [7]—can be seen as an experiment in collectivity at the level of atoms.

Equations (15.8–15.10) highlight the f-Group-related *m*-, *n*-, and *o*-operators.

Hence, Eq. (15.8), f-Group Correlation with Harmony, follows:

$$m\left(X_{\text{atom}}^{(H)}\right) = \text{f-Group} \propto f(\text{Harmony}). \tag{15.8}$$

Equation (15.9), f-Group Set as *n*-Operator Output:

$$n\left(X_{\text{atom}}^{(H)}\right) = \text{f-Group}_{\text{set}} \ni \{\text{Interconnected, Cohesion, Collectivity} \ldots.\}. \tag{15.9}$$

Equation 15.10, f-Group Seed Configuration as o-Operator Output, is the result of the o-operator acting at M1:

$$\text{f-Group}_{\text{seed}} = o\left(X_x^{\{\text{H}\}}\right) = \sum_{x=1}^{N} h\left(a_x X_x^{\{\text{H}\}} + \sum_{y=1}^{M} b_y Y_{y,x}^{\{\text{K, Pr,Po,H}\}}\right). \tag{15.10}$$

Here:

- $o\left(X_x^{\{\text{H}\}}\right)$: o-operator output acting on primary Harmony elements $X_x^{\{\text{H}\}}$.
- $Y_{y,x}^{\{\text{K, Pr,Po,H}\}}$: Secondary contributions Knowledge, Presence, Power, and Harmony.
- $a_x > b_y$: Weighting factors for primary and secondary components.

15.4 s-Group

The s-Group reflects Power through energy release and reactivity. These elements, such as Hydrogen and Helium, Lithium and Sodium, and so on, serve as the energetic pioneers, driving stellar processes [8] and forming the foundational energy systems of matter.

Equations (15.11–15.13) highlight the s-Group-related *m*-, *n*-, and *o*-operators.

Hence, Eq. (15.11), s-Group Correlation with Power, follows:

$$m\left(X_{\text{atom}}^{(\text{Po})}\right) = \text{s-Group} \ \propto f\,(\text{Power}). \tag{15.11}$$

Equation (15.12), s-Group Set as *n*-Operator Output:

$$n\left(X_{\text{atom}}^{(\text{Po})}\right) = \text{s-Group}_{set} \ni \{\text{Energy, Reactivity, Pioneer}\ldots.\}. \tag{15.12}$$

Equation (15.13), s-Group Seed Configuration as *o*-Operator Output, is the result of the *o*-operator acting at M1:

$$\text{s-Group}_{\text{seed}} = o\left(X_x^{(\text{Po})}\right) = \sum_{x=1}^{N} h\left(a_x X_x^{(\text{Po})} + \sum_{y=1}^{M} b_y Y_{y,x}^{(\text{K, Pr,Po,H})}\right). \tag{15.13}$$

Here:

- $o\left(X_x^{(\text{Po})}\right)$: *o*-operator output acting on primary Harmony elements $X_x^{(\text{Po})}$.
- $Y_{y,x}^{(\text{K, Pr,Po,H})}$: Secondary contributions Knowledge, Presence, Power, and Harmony.
- $a_x > b_y$: Weighting factors for primary and secondary components.

15.5 A Unified QIQD Representation

A unified QIQD representation at the level of atoms is depicted by Eq. (15.14), Atom Unified QIQD:

$$Q_{\text{atom}} = \sum_{x \in \{\text{K},\text{Pr},\text{Po},\text{H}\}} \sum_{y \in \{1,\ldots,N\}} c_{x,y}\left(o\left(X_y^{(x)}\right) \wedge n\left(X_y^{(x)}\right) \wedge m\left(X_y^{(x)}\right)\right). \tag{15.14}$$

Here:

- $X_y^{(x)}$: Represents the primary components of Knowledge, Presence, Power, and Harmony.
- o, n, m: Operators corresponding to M1 (seed level), M2 (pure property level), and M3 (complex agglomerate generation level) dynamics.
- $c_{x,y}$: Coefficients reflecting amplitude contributions of each property.

The equation reflects the cumulative contributions of all seeds (y) to the properties (x) of atoms. By tracing each $X_y^{(x)}$ through the layered dynamics of the o-, n-, and m-operators, the equation captures the deterministic organization encoded at the level of atoms.

Note that in QIQD the energetic imprint of each atom itself will encapsulate fundamental properties such as presence, power, knowledge, and harmony, natural to a specific atom itself. These imprints are represented through spectral characteristics, including frequency, amplitude, phase, and polarization. Frequency reflects the vibrational nature of a property (knowledge), amplitude represents its strength (power), phase encodes relational dynamics (harmony), and polarization captures directional orientation (presence), and will be instrumental in considering the architecture and dynamics of qibits that will be further discussed in Chap. 18. Further, molecules, and more complex macromolecules, and by extension cells (or what will be referred to as Cell Equivalents—CE) can also be thought of as having an energetic imprint that will provide insight into its essential set of functions, though these will require more sophisticated technology to perceive and leverage.

15.6 Further Reading

- "The Origin and Nature of Life on Earth: The Emergence of the Fourth Geosphere" by Smith, E., and H. J. Morowitz: Examines how carbon's unique bonding properties underlie life's molecular complexity, making it the elemental bedrock of biological systems.
- "Chemistry of the Elements" by Greenwood, N. N., and Earnshaw, A: A comprehensive guide to all the elements, including detailed coverage of transition metals—discussing their chemical behavior, oxidation states, bonding, and industrial significance.

- "Nature's Building Blocks: An A–Z Guide to the Elements" by Emsley, J: Offers accessible descriptions of each element, detailing historical discovery, applications, and how transition metals such as Iron, Copper, and Nickel provide foundational support for modern industrial and technological advancements.
- "Inorganic Chemistry" by Housecroft, C. E., and Sharpe, A. G. Harlow: Discusses the electronic structure, reactivity, and complex chemistry of the lanthanide and actinide series, including why partially filled f-orbitals can produce "collective" and sometimes "unusual" properties, such as distinctive spectroscopic and magnetic characteristics.
- "Principles of Stellar Evolution and Nucleosynthesis" by Clayton, D. D.: A thorough textbook on how stars evolve and synthesize elements. Outlines the proton–proton chain and the helium fusion process, underscoring how hydrogen and helium lie at the heart of stellar energy release.
- "Mendeleyev's Dream: The Quest for the Elements" by Paul Strathern tells the captivating story of how chemistry evolved from ancient alchemy to the modern science we know today, culminating in Dmitri Mendeleyev's creation of the periodic table.

References

1. Brown TL, LeMay HE Jr, Bursten BE, Murphy CJ, Woodward PM (2018) Chemistry: the central science, 14th edn. Pearson, New York, NY
2. Levine IN (2009) Quantum chemistry, 6th edn. Pearson Prentice Hall, Upper Saddle River, NJ
3. Alberts B, Johnson A, Lewis J, Morgan D, Raff M, Roberts K, Walter P (2014) Molecular biology of the cell, 6th edn. Garland Science, New York
4. Wolf S, Tauber RN (2000) Silicon processing for the VLSI era, vol 1: Process technology. Lattice Press, Sunset Beach, CA
5. Callister WD, Rethwisch DG (2018) Materials science and engineering: an introduction, 10th edn. Wiley, Hoboken, NJ
6. Gschneidner KA, Eyring L (eds) (1978–2000) Handbook on the physics and chemistry of rare earths. North-Holland, Amsterdam
7. Morss LR, Edelstein NM, Fuger J (eds) (2006) The chemistry of the actinide and transactinide elements, 3rd edn. Springer, Dordrecht
8. Burbidge EM, Burbidge GR, Fowler WA, Hoyle F (1957) Synthesis of the elements in stars. Rev Mod Phys 29(4):547–650

Chapter 16
QIQD Dynamics at the Level of Macromolecules in Cells

Abstract This chapter explores a novel interpretation of cellular life through the lens of Quaternary Interpretation of Quantum Dynamics (QIQD). QIQD proposes that four fundamental properties—Knowledge, Presence, Harmony, and Power—manifest at the quantum level and govern the organization and function of biological systems. It is posited that these properties are embodied by the four major classes of cellular macromolecules: nucleic acids, proteins, lipids, and polysaccharides. Using the mathematical framework of QIQD, specific equations that correlate each macromolecule type with its corresponding property are derived. This analysis reveals how the unique characteristics of each macromolecule contribute to the overall resilience, adaptability, and structure of living cells. Nucleic acids, as carriers of genetic information, embody Knowledge. Proteins, the dynamic workhorses of the cell, represent Presence. Lipids, forming boundaries and compartments, embody Harmony. Finally, polysaccharides, serving as energy reservoirs and structural reinforcements, reflect Power. By unifying these individual QIQD representations into a single equation, a novel perspective is offered on the intricate interplay of macromolecules within the cell. This framework provides a new lens for understanding the fundamental principles governing cellular life and opens avenues for further investigation into the quantum underpinnings of biological systems.

At the cellular level, life emerges through the coordinated interplay of four fundamental macromolecule types: nucleic acids, proteins, lipids, and polysaccharides [1]. Within the QIQD framework, these macromolecules reflect the dynamic manifestation of Knowledge, Presence, Harmony, and Power, respectively, core universal properties at the quantum scale.[1] Here, each molecular type acts as an embodiment of one of these properties, ensuring life's resilience, adaptability, and structure and is represented by Eq. (16.1) macromolecule as *m*-Operator output:

---

[1] Initial renderings of this idea appeared across the series of Cosmology of Light books referred to earlier and then subsequently in several IEEE papers—https://ieeexplore.ieee.org/author/37086022058, and here has been further developed as per the overarching quantum-based mathematical framework.

P. Malik, *Pioneering New Avenues in Quantum Technology*, Studies in Smart Technologies, https://doi.org/10.1007/978-981-96-5463-5_16

$$\text{Cellular Object} = m(X_{\text{cell}}) = m(\text{Pr, Po, K, H})$$
$$\rightarrow \{\text{Nucleic Acids, Proteins, Lipids, Polysaccharides}\} \quad (16.1)$$

Here '*m*' refers to the property generator from M3-dynamics. X refers to the archetype of information-energy that embodies the essence of organization as introduced in association with UQO(X) (4.19).

The following sections surface QIQD equations for each of the cellular objects. Finally each of the cellular objects are unified in a single QIQD representation.

16.1 Nucleic Acids

Nucleic acids function as the carriers of genetic information, encoding and transmitting the blueprints that dictate cellular structure and function [2]. Within QIQD, this reflects the property of "Knowledge", depicted by Eq. (16.2), Nucleic Acid Correlation with Knowledge:

$$m\left(X_{\text{cell}}^{(K)}\right) = \text{Nucleic Acids} \propto f(\text{Knowledge}) \quad (16.2)$$

The properties of nucleic acids can be described by the knowledge set, the result of the *n*-operator at M2, and is depicted by Eq. (16.3), Nucleic Acid Set as *n*-Operator Output:

$$n\left(X_{\text{cell}}^{(\text{K})}\right) = \text{Nucleic Acid}_{\text{set}} \ni \{\text{Conceiving law,}$$
$$\text{Spreading knowledge, Adjusting blueprints, } \ldots.\} \quad (16.3)$$

Equation (16.4), Nucleic Acid Seed Configuration as *o*-Operator Output, is the result of the o-operator acting at M1:

$$\text{Nucleic Acid}_{\text{seed}} = o\left(X_x^{(\text{K})}\right) = \sum_{x=1}^{N} h\left(a_x X_x^{(\text{K})} + \sum_{y=1}^{M} b_y Y_{y,x}^{(\text{K, Pr,Po,H})}\right) \quad (16.4)$$

Here:

- $o\left(X_x^{(\text{K})}\right)$: *o*-operator output acting on primary knowledge elements $X_x^{(\text{K})}$.
- $Y_{y,x}^{(\text{K,Pr,Po,H})}$: Secondary contributions Knowledge, Presence, Power, and Harmony.
- $a_x > b_y$: Weighting factors for primary and secondary components.

16.2 Proteins

Proteins are the embodiment of Presence and are the dynamic workhorses of the cell, fulfilling structural, catalytic, and mechanical roles [3]. Within QIQD this depicted by Eq. (16.5), Proteins Correlation with Presence:

$$m\left(X_{\text{cell}}^{(\text{Pr})}\right) = \text{Proteins} \propto f(\text{Presence}) \tag{16.5}$$

The properties of proteins can be described by the presence set, the result of the n-operator at M2, and is depicted by Eq. (16.6), Protein Set as n-Operator Output:

$$n\left(X_{\text{cell}}^{(\text{Pr})}\right) = \text{Protein}_{\text{set}} \ni \{\text{Service, Perfection, Diligence} \ldots .\} \tag{16.6}$$

Equation (16.7), Protein Seed Configuration as o-Operator Output, is the result of the o-operator acting at M1:

$$\text{Protein}_{\text{seed}} = o\left(X_x^{(\text{Pr})}\right) = \sum_{x=1}^{N} h\left(a_x X_x^{(\text{Pr})} + \sum_{y=1}^{M} b_y Y_{y,x}^{(\text{K, Pr,Po,H})}\right) \tag{16.7}$$

Here:

- $o\left(X_x^{(\text{Pr})}\right)$: o-operator output acting on primary Presence elements $X_x^{(Pr)}$.
- $Y_{y,x}^{(\text{K, Pr,Po,H})}$: Secondary contributions Knowledge, Presence, Power, and Harmony.
- $a_x > b_y$: Weighting factors for primary and secondary components.

16.3 Lipids

Lipids form cellular boundaries and promote compartmentalization, fostering cohesion and unity within cells [4]. Within QIQD, lipids embody "Harmony", enabling balance, cooperation, and specialization. Equation (16.8), Lipids Correlation with Harmony, follows:

$$m\left(X_{\text{cell}}^{(\text{H})}\right) = \text{Lipids} \propto f(\text{Harmony}) \tag{16.8}$$

The properties of lipids can be described by the harmony set, the result of the n-operator at M2, and is depicted by Eq. (16.9), Lipid Set as n-Operator Output:

$$n\left(X_{\text{cell}}^{(\text{H})}\right) = \text{Lipid}_{\text{set}} \ni \{\text{Love, Cohesion, Relationship} \ldots .\} \tag{16.9}$$

Equation (16.10), Lipid Seed Configuration as o-Operator Output, is the result of the o-operator acting at M1:

$$\text{Lipid}_{\text{seed}} = o\left(X_x^{\{\text{H}\}}\right) = \sum_{x=1}^{N} h\left(a_x X_x^{\{\text{H}\}} + \sum_{y=1}^{M} b_y Y_{y,x}^{\{\text{K, Pr,Po,H}\}} \right) \tag{16.10}$$

Here:

- $o\left(X_x^{\{\text{H}\}}\right)$: o-operator output acting on primary Harmony elements $X_x^{\{\text{H}\}}$.
- $Y_{y,x}^{\{\text{K, Pr,Po,H}\}}$: Secondary contributions Knowledge, Presence, Power, and Harmony.
- $a_x > b_y$: Weighting factors for primary and secondary components.

16.4 Polysaccharides

Polysaccharides serve as energy reservoirs and structural reinforcements, empowering cells to function and adapt [5]. Within QIQD, polysaccharides reflect the property of "Power", symbolizing energy, resilience, and strength. Equation (16.11), Polysaccharides Correlation with Power, follows:

$$m\left(X_{\text{cell}}^{(\text{Po})}\right) = \text{Polysaccharides} \propto f(\text{Power}) \tag{16.11}$$

The properties of polysaccharides can be described by the Power set, the result of the n-operator at M2, and is depicted by Eq. (16.12), Polysaccharide Set as n-Operator Output:

$$n\left(X_{\text{cell}}^{(\text{Po})}\right) = \text{Polysaccharide}_{\text{set}} \ni \{\text{Courage, Energy, Resilience} \ldots.\} \tag{16.12}$$

Equation (16.13), Polysaccharide Seed Configuration as o-Operator Output, is the result of the o-operator acting at M1:

$$\text{Polysaccharide}_{\text{seed}} = o\left(X_x^{(\text{Po})}\right) = \sum_{x=1}^{N} h\left(a_x X_x^{(\text{Po})} + \sum_{y=1}^{M} b_y Y_{y,x}^{(\text{K, Pr,Po,H})} \right) \tag{16.13}$$

Here:

- $o\left(X_x^{(\text{Po})}\right)$: o-operator output acting on primary Harmony elements $X_x^{(\text{Po})}$.
- $Y_{y,x}^{(\text{K,Pr,Po,H})}$: Secondary contributions Knowledge, Presence, Power, and Harmony.
- $a_x > b_y$: Weighting facto for primary and secondary components.

16.5 Unified QIQD Representation

A unified QIQD representation at the cellular level is depicted by Eq. (16.14), Macromolecular Unified QIQD:

$$Q_{\text{cell}} = \sum_{x \in \{K,Pr,Po,H\}} \sum_{y \in \{1,\ldots,N\}} c_{x,y} \left(o\left(X_y^{(x)}\right) \wedge n\left(X_y^{(x)}\right) \ni m\left(X_y^{(x)}\right) \right) \quad (16.14)$$

Here:

- $X_y^{(x)}$: Represents the primary components of Knowledge, Presence, Power, and Harmony.
- o, n, m: Operators corresponding to M1 (seed level), M2 (pure property level), and M3 (complex agglomerate generation level) dynamics.
- $c_{x,y}$: Coefficients reflecting amplitude contributions of each property.

The equation reflects the cumulative contributions of all seeds (y) to the properties (x) of macromolecules. By tracing each $X_y^{(x)}$ through the layered dynamics of the o-, n-, and m-operators, the equation captures the deterministic organization encoded at the cellular macromolecule level.

16.6 Further Reading

- "Molecular Cell Biology" by Lodish, H., Berk, A., Kaiser, C. A., Krieger, M., Scott, M. P., Bretscher, A., Ploegh, H., & Matsudaira, P.: Covers the molecular composition of cells, including chapters on nucleic acids (hereditary information), proteins (structure and catalysis), lipids (membrane formation and signaling), and carbohydrates (energy storage and cell recognition).
- "Lehninger Principles of Biochemistry" by Nelson, D. L., & Cox, M. M.: Provides a detailed biochemical perspective on how each macromolecule category contributes to metabolism, genetic regulation, membrane dynamics, and energy transformations within living systems.
- "The Origins of Order: Self-Organization and Selection in Evolution" by Kauffman, S. A.: Discusses how emergent complexity arises in biological systems, providing a theoretical foundation for connecting molecular interactions with larger concepts of adaptability and "universal" properties.
- "What Is Life?" by Schrödinger, E.: Early exploration of how quantum principles might intersect with fundamental biological processes.
- "Molecular Quantum Mechanics" by Peter Atkins and Ronald Friedman: This text delves into the application of quantum mechanics to understand molecular behavior
- "Physical Biology of the Cell" by Rob Phillips, Jane Kondev, Julie Theriot, and Hernan Garcia: This book explores the physical and chemical principles underlying cellular processes, offering a valuable perspective on how macromolecules function within the cellular environment
- "Philosophy of Biology" by Elliott Sober: This book delves into the philosophical foundations of biology.
- "Genes XII" by Benjamin Lewin: This comprehensive text on genetics provides a deep dive into the structure and function of nucleic acids.

- "Introduction to Protein Structure" by Carl Branden and John Tooze: This book explores the intricate world of protein structure and folding, offering insights into how proteins carry out their diverse roles.
- "The Lipid Handbook" edited by Frank D. Gunstone, John L. Harwood, and Fred B. Padley: This handbook provides a detailed overview of lipids, their properties, and their functions.
- "Essentials of Glycobiology" edited by Ajit Varki et al.: This book explores the diverse world of polysaccharides, their biological roles, and their significance in energy storage and structural support.

References

1. Alberts B, Johnson A, Lewis J, Morgan D, Raff M, Roberts K, Walter P (2014) Molecular biology of the cell, 6th edn. Garland Science, New York, NY
2. Crick FHC (1970) Central Dogma of molecular biology. Nature 227(5258):561–563
3. Kyte J (2006) Structure in protein chemistry, 2nd edn. Garland Science, New York, NY
4. Singer SJ, Nicolson GL (1972) The fluid mosaic model of the structure of cell membranes. Science 175(4023):720–731
5. Voet D, Voet JG, Pratt CW (2016) Fundamentals of biochemistry: life at the molecular level, 5th edn. John Wiley & Sons, Hoboken, NJ

Part V
Prelude to a QIQD Design

Chapter 17
Key Quantum Computing Principles in QIQD

Abstract The Quaternary Interpretation of Quantum Dynamics (QIQD) offers a novel framework for quantum computing by merging deterministic structures, layered interactions, and higher-order dynamics with the conventional probabilistic underpinnings of quantum theory. In this paradigm, computation arises from "Universal Quantum Objects" (UQOs) that integrate multiple Hilbert spaces and reveal both probabilistic and deterministic aspects across different meta-layers. Central to QIQD is the concept of "qualified determinism," which balances quantum uncertainty with structured causality, thereby extending the domain of quantum entanglement to multiple layers of organization. The framework introduces "seeds" as foundational computational units, encoding space, time, energy, and gravity (STEG) in ways that shape both emergent and deterministic outcomes. "Quantum object code" describes how each quantum entity evolves, suggesting that the uncertainty principle can be reinterpreted as arising from layered dynamics rather than from a strict quantum limitation. A modified Schrödinger equation, incorporating meta-layer interactions and deterministic influences, further refines predictive power. Within QIQD, "qibits" (Quaternary Quantum Bits) store energetic information across layers and enable feedback loops that can induce states of quantum certainty, thereby shifting the computational stratum. The resulting architecture emphasizes specialized hardware for each meta-layer—ranging from atomic-level operations ($M1$) to electromagnetic manipulations ($M2$) and cellular equivalents ($M3$)—while coherence and entanglement drive computational utility across these layers. By favoring the generation of inherent properties over probabilistic measurement outcomes, QIQD broadens the horizons of quantum computing, paving the way for richer and more deterministically guided applications.

The Quaternary Interpretation of Quantum Dynamics (QIQD) provides a transformative framework for quantum computing, shifting from traditional probabilistic models to architectures rooted in deterministic structures, multi-layered interactions, and higher-order dynamics. This paradigm redefines the nature of computation by integrating quantum certainty, seeds of potentiality, and quantum object code scripted in space, time, energy, and gravity (STEG) dimensions. Based on the discussion in

P. Malik, *Pioneering New Avenues in Quantum Technology*, Studies in Smart Technologies, https://doi.org/10.1007/978-981-96-5463-5_17

the preceding chapters several core principles and insights emerge that will inform the design of alternative quantum computing architectures.

These principles are laid out in order, with subsequent principles building on previous principles.

1. The heart of quantum computation is driven by a Universal Quantum Object (UQO):
   a. UQOs reveal that any quantum object comprises several layers. Quantum objects are unified entities that integrate several Hilbert spaces, ranging from a base-layer driven by probabilistic interaction to more deterministic structure at meta-layers.
   b. UQOs also highlight the essential symmetry—based on Pr, Po, K, H—that reveals a different face of itself in each Hilbert space.
   c. The Universal Quantum Object (UQO) framework governs the transformation of X across layers of organization, where X is a unified information-energy archetype that manifests as emergent structures across UQO.
   d. The UQO(X) Eq. (17.1) is reproduced here and will be further unpacked through subsequent principles.

$$\mathrm{UQO}(X) = \begin{bmatrix} |\psi_\infty\rangle = m(X; \mathrm{Pr}, \mathrm{Po}, \mathrm{K}, \mathrm{H}) \\ (\downarrow P_k|\psi_\infty\rangle) \\ |\psi_k\rangle = n(X; S_{\mathrm{Pr}}, S_{\mathrm{Po}}, S_{\mathrm{K}}, S_{\mathrm{H}}) \\ (\downarrow P_g|\psi_k\rangle) \\ |\psi_g\rangle = o(X; S_{\mathrm{Pr}} \times S_{\mathrm{Po}} \times S_{\mathrm{K}} \times S_{\mathrm{H}}) \\ (\downarrow P_{\mathrm{c}}\left|\psi_g\right\rangle) \\ |\psi_c\rangle = p(X; S, T, E, G) \end{bmatrix} \tag{17.1}$$

2. Qualified Determinism Construct (QDC) highlights qualified determinism—the balance between determinism and probability:
   a. QDC reconciles deterministic and probabilistic elements in quantum dynamics. It connects the layered architecture, showing how deterministic structures emerge from quantum systems. This principle ensures that quantum systems can transcend statistical randomness, aligning with emergent behaviors.
   b. Quantum objects are governed by a balance between deterministic structure and quantum uncertainty, termed qualified determinism. This construct bridges multiple layers of causality (B-Layer and meta-layers $M1, M2, M3$), enabling computations that integrate surface-level dynamics with deeper deterministic blueprints.
   c. The equation for QDC (17.2) highlighting the essential driver of layer-based causality is reproduced here.

$$\text{QDC} = \begin{bmatrix} M3 \rightarrow m(\text{Pr, Po, K, H}) \\ \downarrow\uparrow \\ M2 \rightarrow n(S_{\text{Pr}}, S_{\text{Po}}, S_{\text{K}}, S_{\text{H}}) \\ \downarrow\uparrow \\ M1 \rightarrow o(S_{\text{Pr}} \times S_{\text{Po}} \times S_{\text{K}} \times S_{\text{H}}) \\ \downarrow\uparrow \\ B \rightarrow p(S, T, E, G) \end{bmatrix} \tag{17.2}$$

3. There are different types of entanglement at play and that need to be leveraged:
   a. Entanglement is a fundamental concept in quantum mechanics, but here it's extended to span multiple layers. These diverse entanglement types allow the design of tailored quantum operations across layers.
      i. P-Entanglement: Surface-level entanglement tied to base-layer computation.
      ii. O-Entanglement ($M1$): Ensures uniqueness of seeds and manipulates their configurations.
      iii. N-Entanglement ($M2$): Imprints purity of properties into qibits.
      iv. M-Entanglement ($M3$): Reflects higher-order agglomerates, increasing coherence, and influencing the creation of complex systems.
   b. The different types of entanglement in QIQD that must be accounted for in B-Layer operations (17.3) is reproduced here.

$$B - \text{Layer Entanglement}$$
$$= f\begin{pmatrix} \text{Entanglement}_m : M3 \rightarrow \text{complex agglomerate} \\ \downarrow\uparrow \\ \text{Entanglement}_n : M2 \rightarrow \text{pure property} \\ \downarrow\uparrow \\ \text{Entanglement}_o : M1 \rightarrow \text{set of properties} \\ \downarrow\uparrow \\ \text{Entanglement}_p : B \rightarrow \text{constructed} \end{pmatrix} \tag{17.3}$$

4. Seeds are foundational computational objects:
   a. Seeds are central to QIQD, act as carriers of computational information, and provide the mechanism for bridging abstract quantum states with materialized outcomes. Seeds encode specific combinations of STEG dynamics, enabling computations that reflect both deterministic and emergent properties.
   b. Seed-based computation introduces a layer of granularity and functionality, enhancing adaptability in quantum systems and is represented by (17.4) reproduced here.

$$\text{Seed} = \text{Xa} + \sum_{i=0}^{n} Y_i b_i, \quad \text{where}\begin{bmatrix} X, Y \in [S_{\text{Pr}}, S_{\text{Po}}, S_{\text{K}}, S_{\text{H}}] \\ a, b \text{ are integers; } a > b \end{bmatrix}. \tag{17.4}$$

5. Quantum object code gives insight into how a quantum object will behave from the point of view of its properties:
   a. Each quantum object is accompanied by a fourfold STEG code, enabling it to compute its present and future possibilities.
   b. Equations (17.5–17.8) represent the quantum object code for a simple hypothetical atom showing how the initial seed embedded in the equation for space evolution (S_e), is central for understanding the dynamics of time evolution (T_e), energy evolution (E_e), and gravity evolution (G_e) that exist in unison as the four-dimensional code for any quantum object. Θ(seed) in (17.5) ensures that a 'seed' exists as per (17.4).

$$S_e = h\left(\vec{X} + \sum_{i=0}^{n} \vec{Y}_i\right).\Theta(\text{seed}) \tag{17.5}$$

$$T_e = h\left(T_o + \tilde{\vec{X}} + \sum_{i=0}^{n} \tilde{\vec{Y}}_i\right).\Theta(S_e) \tag{17.6}$$

$$E_e = h\left(E_o + \tilde{\vec{X}} + \sum_{i=0}^{n} \tilde{\vec{Y}}_i\right).\Theta(S_e).(1 + \eta.\mathcal{M}(S_k)) \tag{17.7}$$

$$G_e = h\left(G_o + \tilde{\vec{X}} + \sum_{i=0}^{n} \tilde{\vec{Y}}_i\right).\Theta(S_e).\left(1 + \eta.\mathcal{U}\left(\vec{X}, \left\{\vec{Y}_i\right\}\right)\right) \tag{17.8}$$

6. The uncertainty principle is an expression of emergent behavior:
   a. This reinterpretation shifts the quantum paradigm since the uncertainty principle becomes an expression of emergent behavior, not a fundamental limitation.
   b. Deterministic antecedent structures guide the manifestation of quantum phenomena. While there is an effective decrease in uncertainty due to the deterministic meta-level influence and entanglement, there is also a loss of information in the projection of such determinism onto the B-layer.
   c. Eq. (17.9) reproduced here captures such uncertainty at the B-layer.

$$\Delta Q_1 \Delta Q_2 \geq \frac{h}{4\pi} - \left(\sum_k (f_k(M_k) + \xi_k E_k)\right) + (\alpha \text{Dim}_{\text{Loss}} + \beta \Delta S + \gamma \eta + \delta(1 - F)) \tag{17.9}$$

7. Evolution of quantum systems is driven by a modified Schrödinger equation

a. The modified Schrödinger equation accounts for the influence of meta-layers reflecting not just the evolution of a wavefunction ψ at the B-layer (base layer) but also the influence of meta-layers ($M1$, $M2$, and $M3$), which contribute deterministic, pure property, and complex agglomerate structures.
b. Modified Schrödinger dynamics enable more accurate simulations of real-world quantum behaviors.
c. The modified Schrodinger Eq. (17.10) highlighting STEG quantization and meta-layer contributions is reproduced here.

$$i\frac{h}{4\pi}\frac{\partial}{\partial x}\psi_B = \left(H_B + \sum_{k=S,T,E,G} h\left(x_k a_k + \sum_{i=0}^{n} Y_{i,k} b_{i,k}\right)\right)\psi_B + \sum_{i-1}^{3} f_i\left(H_{M_i}\psi_{M_i}\right) \tag{17.10}$$

8. Qibits are fundamental elements that store energetic information and connect different layers together:

a. Qibits can be thought of as ‘Quaternary Quantum Bits,’ and emerge as the computational foundation for multi-layered architectures.
b. Qibits incorporate the energetic imprint of properties and may exist at any layer of the quantum computing architecture.
c. They also connect layers together by acting as a buffer that stores the result of a computation that may have taken place in another layer.
d. Essentially qibits can be captured by Eq. (17.11).

$$\text{qibit} = f(\text{QDC}) \tag{17.11}$$

9. Creating a state of quantum certainty through feedback changes seed dynamics:

a. The possibility of quantum certainty introduces a deterministic component to quantum computing, grounded in feedback loops between meta-layers ($M1$, $M2$, $M3$) and the base layer (B).
b. Quantum certainty introduces a different type of quantum computing referred to as ‘$(W + W\text{o})(L + L\text{o})$’ in Chap. 1.
c. The relevant equations from Chap. 11 on quantum certainty are reproduced here.

$$\text{Seed}_{\text{mfc}} = \left(a\vec{X} + \sum_{i=0}^{n} b_i\vec{Y}_i\right) \cdot \Theta(\mathcal{A}(M1, M2, M3)) \tag{17.12}$$

$$\text{Space}_{\text{quantization}} = h\left(aX + \sum_{i=0}^{n} b_i Y_i\right).\Theta(S_{\text{mfc}}) \tag{17.13}$$

$$\text{Time}_{\text{quantization}} = h\left(aX + \sum_{i=0}^{n} b_i Y_i\right).\Theta(S_{\text{mfc}}) \tag{17.14}$$

$$\text{Energy}_{\text{quantization}} = h\left(aX + \sum_{i=0}^{n} b_i Y_i\right) \cdot \Theta(S_{\text{mfc}}) \tag{17.15}$$

$$\text{Gravity}_{\text{quantization}} = h\left(aX + \sum_{i=0}^{n} b_i Y_i\right) \cdot \Theta(S_{\text{mfc}}) \tag{17.16}$$

10. Continuous STEG quantization will change the computational stratum:
    a. Materialization within quantum systems is governed by the coordinated quantization of space, time, energy, and gravity.
    b. A threshold level of STEG quantizations will alter the computational stratum by changing the state of qibits.
    c. STEG quantization integrates feedback, enabling computational outputs to transition from abstract possibilities to tangible realities. This principle aligns with QIQD's emphasis on the interplay of deterministic and emergent dynamics.
    d. The cumulative change to the computational stratum is envisioned to be bound by (17.17) reproduced below.

$$T_{n+1} = \text{TC}_n\left(\Pi_{z\in\{S,T,E,G\}} Q_Z^{\alpha} \cdot \mathcal{M}\left(\vec{X}, \left\{\vec{Y_i}\right\}\right) \cdot \Theta(\mathcal{A}(M1, M2, M3))\right) \tag{17.17}$$

11. QIQD implies tailored hardware and algorithms:
    a. Different meta-layers require specialized hardware and gates:
       i. $M1$: Atoms as computational objects, acting on qibits to imprint configurations.
       ii. $M2$: Electromagnetic spectrum manipulation for encoding pure properties.
       iii. $M3$: Cells or Cellular Equivalents (CE) as carriers of complex agglomerates (will be discussed further in Chap. 18).
    b. These modular, multi-layered designs facilitate advanced quantum computing architectures.
12. Quantum utility and coherence will change as the underlying architecture changes:
    a. Utility of quantum systems increases with coherence (C) and entanglement.
    b. Coherence (C) drives superposition and the ability to explore Hilbert space. High coherence enhances computational power and sensing capabilities.
    c. Different degrees of coherence are captured by (17.18) reproduced below.

$$
\begin{aligned}
&\text{Quantum Utility} \\
&= \begin{bmatrix} M3 \rightarrow m(\mathrm{Pr}, \mathrm{Po}, \mathrm{K}, \mathrm{H}) \\ \downarrow\uparrow \\ M2 \rightarrow n(S_{\mathrm{Pr}}, S_{\mathrm{Po}}, S_{\mathrm{K}}, S_{\mathrm{H}}) \\ \downarrow\uparrow \\ M1 \rightarrow o(S_{\mathrm{Pr}} \times S_{\mathrm{Po}} \times S_{\mathrm{K}} \times S_{\mathrm{H}}) \\ \downarrow\uparrow \\ B \rightarrow p(S, T, E, G) \end{bmatrix} \\
&\rightarrow C\left(\rho_{M_3}\right) > C\left(\rho_{M_2}\right) > C\left(\rho_{M_1}\right) > C(\rho_{\mathrm{B}})
\end{aligned}
\tag{17.18}
$$

Traditional quantum systems rely heavily on measurement-driven paradigms. QIQD shifts the focus to computations that generate inherent qualities and properties, emphasizing deterministic structures over probabilistic outputs. This approach expands the scope of quantum computing to include richer, more meaningful outcomes.

17.1 Further Reading

- "Quantum Theory, the Church–Turing Principle and the Universal Quantum Computer." by Deutsch, D.: David Deutsch's classic paper establishing the theoretical basis for universal quantum computation, extending the Church–Turing principle into the quantum realm.
- "Advanced Synergetics: Instability Hierarchies of Self-Organizing Systems and Devices" by Haken, H.: Explores multi-level dynamics in self-organizing systems. Offers theoretical parallels to QIQD's notion of base-layer and meta-layer interactions guiding emergent structures and coherence.
- "Bell Nonlocality." By Brunner, N., Cavalcanti, D., Pironio, S., Scarani, V., & Wehner, S .: A comprehensive overview of entanglement and nonlocality, laying a foundation for understanding how QIQD distinguishes entanglement types (P-, O-, N-, M-Entanglement) across different layers.
- "Quantum Measurement and Control" by Wiseman, H. M., & Milburn, G. J: Explores measurement-based feedback and control mechanisms in quantum systems, aligning with QIQD's notion of deterministic feedback loops in quantum certainty states.
- "Quantum Computing for Computer Scientists" by Noson S. Yanofsky and Mirco A. Mannucci: This book offers a comprehensive introduction to quantum computing, covering fundamental concepts like qubits, quantum gates, and quantum algorithms.
- "The Hidden Reality: Parallel Universes and the Deep Laws of the Cosmos" by Brian Greene: This book explores different interpretations of quantum mechanics, including those that incorporate elements of determinism. It helps to place QIQD

within the broader landscape of interpretations that grapple with the probabilistic nature of quantum mechanics.

- "Einstein, Bohr, and the Great Debate About the Nature of Reality" by Manjit Kumar: This book delves into the historical debate between Einstein and Bohr about the foundations of quantum mechanics, highlighting different perspectives on determinism and uncertainty. It provides valuable context for understanding the philosophical implications of QIQD's "qualified determinism."
- "Complexity: A Guided Tour" by Melanie Mitchell: This book explores the concept of emergence in complex systems, where higher-level properties arise from interactions at lower levels. This is relevant to QIQD's emphasis on layered interactions and the emergence of deterministic outcomes from quantum uncertainty.
- "Quantum Enigma: Physics Encounters Consciousness" by Bruce Rosenblum and Fred Kuttner: This book delves into the philosophical implications of quantum mechanics, including the role of the observer and the nature of consciousness.

Chapter 18
Envisioning Generators and Qibits

Abstract This chapter explores the conceptual basis for a novel quantum computing hardware architecture based on the Quaternary Interpretation of Quantum Dynamics (QIQD). QIQD seeks to manipulate and compute with the inherent properties of quantum objects, enabling the creation and manipulation of new properties. The proposed hardware consists of three main categories: (1) Generators, including Pure Atom Rings (PARs), Pure Property Chips (PPCs), Pure Property Quantum Dots (PPQDs), and Mixed Atom Rings (MARs), each designed to generate specific dynamics at different meta-layers (M1, M2, M3); (2) Property Storage Units, such as qibits and STEG-Fabric, which store individual or "seed" properties; and (3) Gates, including PP-Adders, PP/PAR-Adders, PP/PAR-Subtracters, and Magnify Gates, which manipulate and combine properties. This chapter focuses on the generators and property storage units, providing an examination of their theoretical functionalities and potential implementations. By combining these elements, QIQD aims to construct algorithms that operate directly on the properties of quantum objects, opening new avenues for quantum computation.

The basic objective of QIQD hardware is to be able to manipulate and compute at the level of properties and functions that naturally exist in quantum objects and conglomerates of quantum objects, and further, to imprint or store such properties, and create new ones.

The hardware envisioned to achieve this may not exist, and in this chapter a conceptual basis is suggested to imagine what such hardware may be. Broad categories of QIQD hardware are envisioned to be the following:

1. Generators (G-M1, G-M2, G-M3) of M1, M2, and M3 dynamics:
    a. These are elements such as a Pure Atom Ring (PAR) used to generate M1 dynamics (G-M1).
    b. Pure Property Chip (PPC) envisioned to be a combination of generators and gates that act together to create single "functional" properties (e.g.,--the energetic imprint of 'malleability'), also at the M1 level (G-M1).

P. Malik, *Pioneering New Avenues in Quantum Technology*, Studies in Smart Technologies, https://doi.org/10.1007/978-981-96-5463-5_18

c. Pure Property Quantum Dot (PPQD) envisioned to create a pure property used to generate M2 dynamics (G-M2).
d. Mixed Atom Ring (MAR) and Cellular Equivalents (CE) envisioned to generate the more complex M3 dynamic (G-M3).
e. G-M3 elements are also envisioned to initiate a computational process by putting the QIQD Processing Unit (QIQD-PU) into a highly coherent state.

2. Property Storage Units:

   a. Qibits which store individual or seed properties.
   b. STEG-Fabric envisioned to contain an array of qibits.

3. Gates manipulate properties by adding, subtracting, and storing them in qibits and the STEG-Fabric:

   a. PP-Adder: Adds pure properties together.
   b. PP/PAR-Adder: Adds pure properties and properties in a PAR together.
   c. PP/PAR-Subtractor: Subtracts a pure property from properties in a PAR.
   d. Magnify Gates: Practically an infinite number of different gates made from combinations of unique photons and PARs or MARs, aimed at magnifying sets of properties.

This chapter will look specifically at Generators—PARs, PPCs, PPQDs, and MARs, and Qibits, the property storage unit. Note that combinations of elements from the preceding categories will create QIQD algorithms [1].

18.1 Pure Atom Ring (PAR)

The Pure Atom Ring (PAR) is depicted in Fig. 18.1, *Pure Atom Ring*:

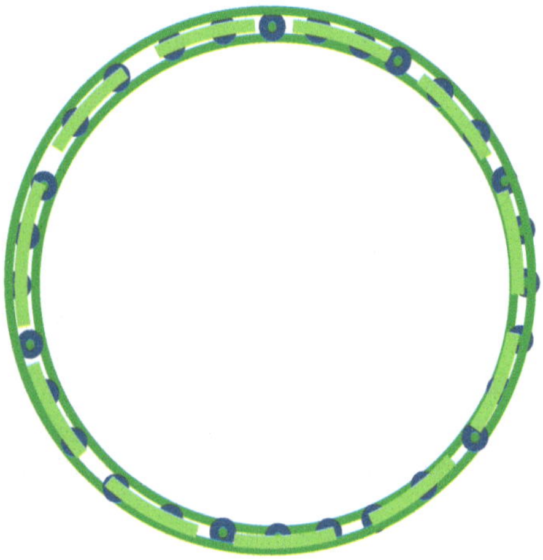

Fig. 18.1 Pure atom ring

To realize the Pure Atom Ring (PAR) as envisioned—where the properties embodied by the atoms in the ring energetically manifest within the circumscribed area—requires integrating advanced techniques from quantum optics, atomic manipulation [2], and nanotechnology. The objective is to create a quantum ring where atoms of a single type are confined to a ring-shaped channel, allowing their collective properties to influence the quantum field within the ring's circumscribed area. This requires a mechanism to confine and stabilize atoms in a ring-shaped configuration and maintain quantum coherence over time. It also necessitates the ability to extract and amplify the properties embodied by the atoms into a quantum field within the ring.

The containment of atoms in a ring may be achieved by engaging complementary technologies that work in unison. The first step is to create a ring-shaped optical trap [3] using a tightly focused laser beam to generate a hollow light ring that traps and confines atoms within its circumference [4]. Once the desired atomic configuration is achieved, the atoms can be embedded in nano-fabricated channels to lock in the ring's structure [5]. The nano-fabricated channels provide a physical substrate that permanently fixes the configuration of the atoms in the ring [6]. Such a channel can also mitigate thermal noise, a significant source of decoherence in quantum systems.

The ring can also be surrounded with a superconducting quantum structure to enable coupling with external magnetic fields [7]. This will enable magnetic field tuning to influence the properties of the trapped atoms and the energetic imprint they create. Further, a superconducting quantum ring creates a quantum-coherent structure that enhances interactions between the atoms in the ring, which coupled with reduction of environmental noise and decoherence effects, ensures the longevity of the PAR's quantum state.

Amplifying the energetic properties of the Pure Atom Ring may involve embedding quantum amplifiers within the ring to enhance the interaction between atomic properties and quantum fields [8]. The collective excitation of atoms, through mechanisms like Rydberg states or Bose–Einstein condensation, enables the extraction and amplification of atomic properties by exciting them to specific energy levels that project their influence into the circumscribed area [9]. Additionally, resonant circuits can be employed to create harmonic coupling, where the atomic configuration imprints its properties harmonically into the surrounding area, further amplifying the ring's energetic impact.

The integration of the Pure Atom Ring (PAR) with the computational framework involves encoding the energetic imprint of the ring's atoms as a quantum field in the circumscribed area, representing their combined properties for computation directly on the field's parameters. This field is further coupled to the STEG-Fabric, allowing the imprint to be dynamically stored and manipulated as part of broader computational processes. Additionally, the energetic imprint serves as an input for QIQD gates, such as PP-Adders and PP/PAR-subtractors, enabling real-time manipulation and processing of quantum properties within the framework.

18.2 Pure Property Chip (PPC)

The Pure Property Chip (PPC) also serves as a G-M1 designed to generate single, though more complex functional properties. These properties can be thought of as belonging to atoms. For example, we know that silver is malleable, ductile, non-corrosive, and so on. Each of these is a single, though more complex functional property.

By bringing together a range of generators and gates, discussed in this chapter, it will be possible to isolate the energetic imprint of such a property. A PPC will house the combination of such generators, qibits, and gates, to create an exclusive property X. Any such chip, PPC_X, where X refers to the exclusive property can now be available to be leveraged in the creation of programmable matter (to be discussed in Chap. 20). Note further, that PPC_X becomes the equivalent of one type of QIQD algorithm.

18.3 Pure Property Quantum Dot (PPQD)

The Pure Property Quantum Dot (PPQD) serves as a G-M2, designed to isolate and generate purer functional properties. In the Quantum Interpretation of Quantum Dynamics (QIQD) framework, quantum dots are engineered as vital components, each meticulously designed to resonate with a unique wavelength of light that corresponds to specific quantum properties. These semiconductor particles, known as quantum dots [10], are nano-sized and exhibit distinct energy levels due to their quantum mechanical behaviors. By carefully adjusting the size, shape, and material composition of these dots, they can be tuned to absorb or emit photons at precise wavelengths, thereby encoding specific quantum properties within their structure [11].

When a quantum dot absorbs a photon of its resonant wavelength, it triggers the excitation of an electron to a higher energy state, effectively embedding the associated quantum property into the dot. The return of the electron to its ground state is marked by the emission of a photon at the same wavelength, signaling the presence of this quantum property. This ability to absorb and emit light at predetermined wavelengths allows quantum dots to function as dynamic processors and storages of quantum information, adapting to the computational needs of the QIQD system.

The functionality of quantum dots extends beyond static encoding. They are capable of dynamic property manipulation through the application of external stimuli, such as electric and magnetic fields or additional light sources. This dynamic adjustment of their energy levels enables the quantum dots to switch between different quantum states or properties rapidly.

Quantum dots are integral to the quantum computing architecture in QIQD [12], where they can be arranged in networks or lattices like the STEG-Fabric. Each dot's specific interaction with certain wavelengths allows for the construction of elaborate computations that involve multiple quantum properties. These interactions are facilitated through the light they emit or absorb, creating a coherent and synchronized system where properties can be transferred seamlessly between different quantum dots or other components of the computing system.

In this way, quantum dots in the QIQD framework are not merely passive elements but are active and adaptable units that encode, manipulate, and relay quantum information, facilitating sophisticated and dynamic interactions within advanced quantum computing architectures.

18.4 Mixed Atom Ring (MAR)

The concept of Mixed Atom Rings (MARs) expands upon the foundational idea of Pure Atom Rings (PARs) by envisioning them as several concentric rings. The Mixed Atom Ring (MAR) is depicted in Fig. 18.2, *Mixed Atom Ring*. In this depiction each ring is representative of a PAR:

Each concentric PAR in a MAR operates independently while being influenced by its neighboring rings. The configuration of these rings can be tuned to specific properties or energetic states, promoting complementary dynamics. Additionally, the radius of each ring is made dynamically adjustable, allowing precise modulation of inter-ring spacing and interaction strength. This capability enables fine-tuning of emergent dynamics and the generation of specific properties.

Further, the assignment of roles to individual rings can enhance functionality. Inner rings may seed fundamental configurations (M1), middle rings may encode pure properties (M2), and outer rings may focus on relational dynamics between these configurations and properties (M3). Together, these layers contribute to the creation of a superimposed energetic field within the area circumscribed by the MAR. This field becomes a reflection of the collective contributions of all concentric

Fig. 18.2 Mixed atom ring

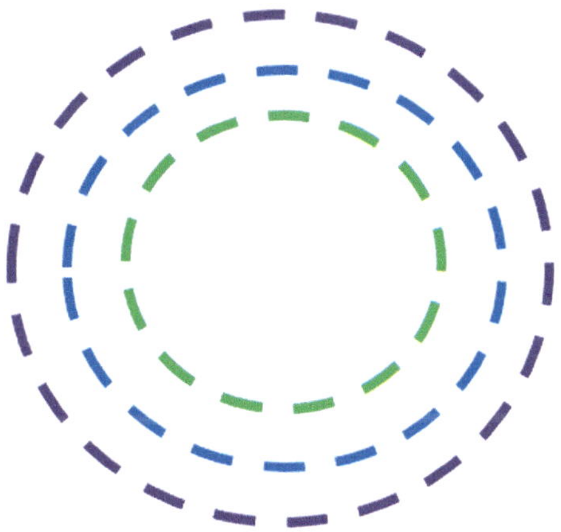

rings, encapsulating the emergent properties of the entire system and can be used to initialize the computation system.

Distinct frequencies or energy levels assigned to each ring allow for controlled interaction patterns. For instance, slightly detuned frequencies can create constructive interference that strengthens specific properties. Furthermore, MARs enable the dynamic mixing of properties across rings, producing unique combinations. An inner ring encoding a "presence"-based property, can be combined with middle rings that due to their focus on pure property generation, can amplify that aspect in the seed configuration represented by the presence-based inner ring. The outer ring that may have a complex signature could increase the coherence of the overall computation due to superposition and entanglement created by it.

The use of optical or magnetic multi-ring systems provides the structural basis for confining atoms or ions in distinct yet interacting rings. Field mediators, such as photonic or electromagnetic fields, can facilitate interactions between the rings, while dynamic coupling gates selectively enhance or suppress specific inter-ring interactions. Feedback systems equipped with real-time sensors can potentially continuously monitor the inter-ring dynamics, adjusting parameters such as spacing and energy levels to optimize emergent properties.

18.5 Property Storage Unit: Qibits

In the framework of QIQD computation, qibits serve as specialized storage units for functional properties, ensuring their coherence, granularity, and dynamic adaptability. These units embody the essence of quantum properties through an encoding process that aligns with the STEG framework. By leveraging spectral modulation, it is envisioned that qibits can store energetic imprints with precision and coherence, making them integral to advanced quantum systems.

Energetic imprints, as suggested in Chap. 15, encapsulate fundamental properties such as presence, power, knowledge, and harmony, which naturally emerge in QIQD quantum systems. These imprints are represented through spectral characteristics, including frequency, amplitude, phase, and polarization. Frequency reflects the vibrational nature of a property (knowledge), amplitude represents its strength (power), phase encodes relational dynamics (harmony), and polarization captures directional orientation (presence). By mapping these spectral features to energetic imprints, qibits provide a dynamic and coherent method for property storage.

The encoding process begins with initializing a qibit in a known quantum state, such as a superposition state. Spectral modulation is then applied to imprint the desired energetic characteristics onto the qibit [13–16]. Frequency modulation adjusts the oscillation frequency to represent dynamic aspects of the property, while amplitude modulation varies the state amplitude to encode magnitude. Phase modulation introduces shifts to represent relational or directional nuances, and polarization modulation reflects spatial or contextual factors. The resulting qibit state incorporates all these spectral components, creating a multi-dimensional representation of the energetic imprint.

Technological advancements make this encoding process feasible. Electro-optic modulators (EOMs) and acousto-optic modulators (AOMs) [17] facilitate high-speed, precise spectral adjustments, allowing for dynamic property representation. Hybrid qubit architectures, such as photonic-spin or superconducting-photonic qubits [7], support the encoding process by integrating spectral features like frequency, amplitude, phase, and polarization into a unified quantum state [18] and can be leveraged in qibit architectures. For example, a photonic qubit might encode frequency and polarization, while a spin qubit encodes phase and amplitude, creating a robust and flexible storage unit. This method of using spectral modulation to encode energetic imprints into qibits offers several advantages. It provides a precise, high-resolution representation of properties, where each spectral component corresponds to a specific aspect of the energetic imprint. The dynamic adjustability of modulation allows qibits to evolve with the computation, adapting to changing properties or imprints.

For instance, consider a workflow where an energetic imprint representing "power" is extracted from a Pure Atom Ring (PAR) or Mixed Atom Ring (MAR). This imprint is spectrally decomposed to identify its frequency, amplitude, polarization, and phase characteristics. An unmodulated qibit is then prepared, and spectral modulation units apply these characteristics to the qibit's state, embedding the imprint within the quantum system. This process transforms the qibit into a coherent storage unit capable of dynamically adapting and interacting with other quantum entities in the system.

Equations (18.1)–(18.4) capture the theoretical essence of how qibits operate within the QIQD framework, providing a quantifiable model that relates to the spectral modulation processes required to manipulate and store quantum properties dynamically.

Equation 18.1, *Qibit Encoding*, is as follows:

$$|\psi\rangle = \sum_{i=1}^{N} c_i |f_i, a_i, \phi_i, p_i\rangle \tag{18.1}$$

Here

- $|\psi\rangle$ is the quantum state of the qibit.
- c_i are coefficients representing the strength of each property.
- f_i, a_i, ϕ_i, p_i represent the frequency, amplitude, phase, polarization of the spectral components respectively.
- N is the number of properties encoded in the qibit.

Equation 18.2, *Dynamic Modulation of Spectral Components*, is as follows:

$$\Delta|\psi\rangle = \sum_{i=1}^{N} \delta c_i |f_i + \Delta f_i, a_i + \Delta a_i, \phi_i + \Delta\phi_i, p_i + \Delta p_i\rangle \tag{18.2}$$

Here, $\Delta f_i, \Delta a_i, \Delta \phi_i, \Delta p_i$ represent changes in the spectral components due to external modulation or internal system dynamics.

Equation 18.3, *Qibit Coherence*, is as follows:

$$C(t) = \frac{1}{N} \sum_{ij} \left| \langle f_i, a_i, \phi_i, p_i | f_j, a_j, \phi_j, p_j \rangle \right| \tag{18.3}$$

Here, $C(t)$ is the coherence measure of the qibit state at time t, capturing the overlap between different spectral states within the qibit.

Equation 18.4, *Entanglement and Interaction Between Qibits*, is as follows:

$$E_{ij} = -Tr\left(\rho_{ij} log \rho_{ij}\right) \tag{18.4}$$

Here, ρ_{ij} is the reduced density matrix for qibits i and j, showing the degree of entanglement between them.

18.6 Further Reading

- "Quantum Algorithm Implementations for Beginners" by Ryan LaRose: This online resource offers a practical guide to implementing quantum algorithms using quantum gates and circuits.
- "Qiskit Textbook": This online textbook (hosted by IBM Quantum) provides a comprehensive introduction to quantum computing using the Qiskit framework. It includes detailed explanations of quantum gates, circuits, and algorithms, with interactive code examples and exercises.
- "Quantum Algorithms for Algebraic Problems" by Andrew Childs and Wim van Dam: This review article provides a comprehensive overview of quantum algorithms for algebraic problems, including detailed explanations of how these algorithms are constructed using quantum gates and circuits.
- "The Physical Implementation of Quantum Computation." by Divincenzo, David P.: Discusses hardware requirements for qubits—coherence, scalability, gating—applicable to any advanced quantum architecture.
- "Building logical qubits in a superconducting quantum computing system" by Gambetta, Chow, and Steffen: This research article provides insights into the complexities of building robust qubits for quantum computation, which is relevant to the challenges of realizing QIQD hardware.
- "Quantum Optics" by Marlan O. Scully and M. Suhail Zubairy: This comprehensive text explores the interaction of light and matter at the quantum level, covering topics such as atom-photon interactions, lasers, and quantum coherence. This is relevant to understanding the principles behind PARs and PPQDs.

- "Nanotechnology: Understanding Small Systems" by Ben Rogers, Jesse Adams, and Sumita Pennathur: This book provides a comprehensive introduction to nanotechnology, including the fabrication and properties of nanostructures like quantum dots.
- "Quantum Dot Devices" by D. Bimberg, M. Grundmann, and N. N. Ledentsov: This book chapter delves into the physics and applications of quantum dots, which are central to the proposed PPQD component of QIQD hardware.
- "Quantum control of light using electromagnetically induced transparency" by Harris: This article introduces the concept of electromagnetically induced transparency (EIT), a technique for manipulating light-matter interactions that could be relevant to spectral modulation in QIQD.
- "Coherent control of quantum dynamics: The dream is alive" by Warren S. Warren, Herschel Rabitz, and Mohammed Dahleh: This article discusses the broader field of coherent control in quantum systems, which is relevant to the precise manipulation of quantum states required for QIQD.
- "Programmable matter" by Seth Goldstein and Todd Mowry: This article provides an overview of the concept of programmable matter, where materials can change their physical properties on demand.

References

1. Barenco A, Bennett CH, Cleve R, DiVincenzo DP, Margolus N, Shor P, Sleator T, Smolin JA, Weinfurter H (1995) Elementary gates for quantum computation. Phys Rev A 52(5):3457–3467
2. Focus on atomtronics-enabled quantum technologies. New J Phys 19:020201 (2017). https://doi.org/10.1088/1367-2630/aa5a6d
3. Morizot, O., Y. Colombe, V. Lorent, H. Perrin, and B. M. Garraway. "Ring Trap for Ultracold Atoms." Physical Review A 74, no. 2 (2006): 023617. https://doi.org/10.1103/PhysRevA.74.023617.
4. Gupta S, Murch KW, Moore KL, Purdy TP, Stamper-Kurn DM (2005) Bose–Einstein condensation in a circular waveguide. Phys Rev Lett 95:143201
5. Reichel J, Vladan V (eds) (2011) Atom chips. Wiley, Weinheim
6. Folman R, Krüger P, Schmiedmayer J, Denschlag J, Henkel C (2002) Microscopic atom optics: from wires to an atom chip. Adv At Mol Opt Phy 48:263–356. https://doi.org/10.1016/S1049-250X(02)80011-8
7. Devoret MH, Schoelkopf RJ (2013) Superconducting circuits for quantum information: an outlook. Science 339(6124):1169–1174. https://doi.org/10.1126/science.1231930
8. Castellanos-Beltran MA, Lehnert KW (2007) Widely tunable parametric amplifier based on a superconducting quantum interference device array resonator. Appl Phys Lett 91(8):083509. https://doi.org/10.1063/1.2773988
9. Weimer H, Müller M, Lesanovsky I, Zoller P, Büchler HP (2010) A rydberg quantum simulator. Nat Phys 6:382–88. https://doi.org/10.1038/nphys1614
10. Michler P (ed) (2003) Single quantum dots: fundamentals, applications and new concepts. Springer, Berlin
11. Semiconductor clusters, nanocrystals, and quantum dots. Science 271(5251):933–937 (1996). https://doi.org/10.1126/science.271.5251.933
12. Quantum computation with quantum dots. Phys Rev A 57(1):120–26 (1998). https://doi.org/10.1103/PhysRevA.57.120

13. Lukens JM, Lougovski P (2017) Frequency-encoded photonic qubits for scalable quantum information processing. Optica 4(1):8–16. https://doi.org/10.1364/OPTICA.4.000008
14. Kwiat PG, Weinfurter H, Herzog T, Zeilinger A, Kasevich MA (1995) Interaction-free measurement. Phys Rev Lett 74(24):4763–4766. https://doi.org/10.1103/PhysRevLett.74.4763
15. Barreiro JT, Wei T-C, Kwiat PG (2008) Beating the channel capacity limit for linear photonic superdense coding. Nat Phys 4:282–286. https://doi.org/10.1038/nphys919
16. Erhard M, Fickler R, Krenn M, Zeilinger A (2018) Twisted photons: new quantum perspectives in high dimensions. Light Sci Appl 7:17146. https://doi.org/10.1038/lsa.2017.146
17. Yariv A, Yeh P (2007) Photonics: optical electronics in modern communications, 6th edn. Oxford University Press, Oxford
18. Ladd TD, Jelezko F, Laflamme R, Nakamura Y, Monroe C, O'Brien JL (2010) Quantum computers. Nature 464:45–53. https://doi.org/10.1038/nature08812

Chapter 19
Envisioning Other QIQD Hardware

Abstract This chapter delves into the conceptual hardware architecture of the Quaternary Interpretation of Quantum Dynamics (QIQD), focusing on property processing units, gates, and the QIQD Processing Unit (QIQD-PU). Central to QIQD hardware is the STEG-Fabric, a dynamic matrix that integrates encoded qibits, each imbued with "energetic imprints" derived from Pure Atom Rings (PARs) and Mixed Atom Rings (MARs). This self-organizing fabric adapts its topology to optimize computational efficiency, enhancing connectivity for active qibits and modulating the network based on computational demands. QIQD gates manipulate these energetic imprints. The PP-Adder combines pure properties, encoding the resultant composite property into a qibit or the STEG-Fabric. The PP/PAR-Subtractor gate, utilizing wavefunction isolation guides and frequency-adjustable circuits, subtracts pure properties from those embedded in a PAR. These gates, through precise manipulation of energetic imprints, embody the core principles of QIQD computation. The QIQD-PU serves as the central computational hub, orchestrating operations across multiple quantum dynamics layers (G-M1, G-M2, G-M3). Its Adaptive Hamiltonian Controller dynamically adjusts the energy landscape based on inputs from property generators, qibit interactions, and feedback from quantum sensors within the STEG-Fabric. This ensures computations remain coherent, stable, and responsive, highlighting the QIQD framework's capacity to redefine computation by operating directly at the level of quantum properties and their interactions.

19.1 Property Processing Unit—STEG-Fabric

The STEG-Fabric in the Quantum Interpretation of Quantum Dynamics (QIQD) framework plays a pivotal role in the integration and functionality of quantum computation. This fabric acts as a dynamic matrix for encoded qibits, each embedded with an energetic imprint derived from Pure Atom Rings (PARs) or Mixed Atom Rings (MARs). Once a qibit is encoded, it is intricately woven into the fabric, making the STEG-Fabric a composite entity that mirrors the cumulative effects of the QIQD computational processes. Additionally, this fabric is not just a passive repository

P. Malik, *Pioneering New Avenues in Quantum Technology*, Studies in Smart Technologies, https://doi.org/10.1007/978-981-96-5463-5_19

but is designed to be self-organizing, a feature that underpins its ability to reflect emergent dynamics vividly.

Central to the STEG-Fabric's operation is its envisioned adaptive topology control [1]. This feature enables the lattice structure of the fabric to dynamically adjust its connectivity based on the ongoing computational demands. For instance, if certain qibits become pivotal in processing or encoding specific properties, the fabric's topology is modified to enhance these connections, thereby elevating the roles of these active qibits. In contrast, qibits that are less active might see a reduction in their connectivity, effectively modulating the network to optimize computational efficiency and responsiveness.

Equation 19.1, *Adaptive Topology Adjustment,* is as shown and will be instrumental modulating the network:

$$C_{ij}(t) = \alpha_{ij}(t).e\left(-\frac{\left|\vec{r}_i(t) - \vec{r}_j(t)\right|}{\xi}\right) \tag{19.1}$$

Here:

- $C_{ij}(t)$ represents the strength of connectivity between qibits i and j at time t.
- $\alpha_{ij}(t)$ is a factor adjusting for the activity level of the qibits.
- $\vec{r}_i(t)$ and $\vec{r}_j(t)$ are their positions in the lattice.
- ξ is a characteristic length scale that modulates connectivity based on qibit proximity and activity.

Another essential envisioned attribute of the STEG-Fabric is its decoherence suppression mechanisms. Quantum states, inherently delicate, are prone to errors from environmental interferences and internal instabilities. The fabric's design with its dynamic reconfigurations will likely not only reduce noise and interference but also stabilize the quantum states within. This is crucial for maintaining the integrity and coherence of the computations handled by the fabric.

Equation 19.2, *Decoherence Suppression Mechanism*, is as shown:

$$\Gamma_{\text{decoh}}(i, t) = \varsigma_{\text{o}}.e\left(-\beta \sum_{j \neq i} C_{ij}(t)\right) \tag{19.2}$$

Here:

- $\Gamma_{\text{decoh}}(i, t)$ quantifies the rate of decoherence for qibit i at time t.
- ς_{o} is the base decoherence rate in isolation.
- β modulates the decoherence suppression effect due to connectivity with other qibits.

Through its lattice design, adaptive topology, and decoherence suppression mechanisms, the STEG-Fabric serves as a foundational component in the architecture of quantum computation [2]. It transforms individual qibits into integrated elements of an adaptive computational system. The fabric is conceived as an active and continuously evolving entity, epitomizing the principles of adaptability, coherence, and efficiency that are central to the advancement of quantum technology.

19.2 Gates

At the core of this system lies the PP-Adder, a gate designed to combine two pure properties into a new composite property. By integrating spectral modulation units, the PP-Adder encodes the resultant property into a qibit or the STEG-Fabric, allowing it to become a functional part of broader computations. This encoding process ensures that the composite property maintains coherence and reflects the nuanced interactions between the original inputs.

Equation 19.3, *PP-Adder Functionality*, is as shown:

$$P_{\text{composite}} = \text{Modulate}(\alpha P_1(f_1, a_1, \phi_1, p_1) + \beta P_2(f_2, a_2, \phi_2, p_2)) \tag{19.3}$$

Here:

- $P_{\text{composite}}$ is the new composite property encoded into the qibit or STEG-Fabric.
- $P_1(f_1, a_1, \phi_1, p_1)$ and $P_2(f_2, a_2, \phi_2, p_2)$ represent the spectral characteristics of the two input properties.
- α and β are coefficients that weight the contributions of P_1 and P_2 to the composite property.
- 'Modulate' is a function that applies spectral modulation to the combined properties by adjusting the resulting frequency, amplitude, phase, and polarization to optimize coherence and interaction with the quantum system.

Building on this concept, the PP/PAR-Adder gate extends the functionality by combining pure properties (PP) with those embedded within a Pure Atom Ring (PAR). The addition process is supported by dynamic field coupling, a real-time mechanism that ensures the energetic fields of the combined properties are balanced and coherent [3]. Dynamic field coupling detects and adjusts parameters such as amplitude, phase, and frequency to harmonize the contributions from the pure properties and the PAR. For instance, if a property within the PAR has a stronger imprint, the coupling mechanism adjusts its amplitude to prevent dominance and achieve a balanced output. This process ensures that the resultant composite property accurately reflects the intended combination of inputs while preserving their functional integrity.

Dynamic field coupling operates through an adaptive feedback system, where quantum sensors continuously monitor the properties' spectral characteristics. These sensors provide real-time data, allowing the system to modulate interaction parameters dynamically [4]. For example, if the detected phase of a PAR property is misaligned with the pure property, the coupling gates align the phases to achieve coherent superposition. This adaptability not only balances the energetic contributions but also enables the generation of new, emergent properties that exhibit enhanced functionality.

Equation 19.4, *PP/PAR-Adder Functionality*, is as shown:

$$P_{\text{composite}} = \text{Couple}(\alpha P_{\text{PP}}(f_{\text{PP}}, a_{\text{PP}}, \phi_{\text{PP}}, p_{\text{PP}}) + \beta P_{\text{PAR}}(f_{\text{PAR}}, a_{\text{PAR}}, \phi_{\text{PAR}}, p_{\text{PAR}}), \Gamma) \tag{19.4}$$

Here:

- $P_{\text{composite}}$ is the resultant composite property encoded into the qibit or STEG-Fabric.
- α and β are coefficients representing the relative contributions of the pure property and PAR property to the composite property.
- Γ represents a set of modulation parameters derived from real-time sensor data, including adjustments for amplitude, phase, and frequency to balance the outputs.
- 'Couple' is a function that applies the dynamic field coupling, modifying the combined properties to achieve coherence and balance.

The PP/PAR-Subtractor gate is another vital component, designed to subtract pure properties from those embedded in a PAR. This operation relies on wavefunction isolation guides and frequency-adjustable circuits to achieve precision. The wavefunction isolation guides isolated specific components of the quantum state associated with the property to be subtracted, ensuring that other properties remain undisturbed. This isolation is achieved through advanced technologies such as optical filters or quantum dots, which separate the targeted property with minimal interference [5].

Once isolation is achieved, frequency-adjustable circuits generate a counter-wavefunction that precisely matches the isolated property's spectral signature. This counter-wave, having the same amplitude but an opposite phase, interacts with the isolated property to cancel it out through destructive interference [6]. For example, if a "presence" property encoded at 10 THz needs to be subtracted, the circuit generates a 10 THz counter-wave to neutralize its influence. After subtraction, the remaining wavefunction is reintegrated into the quantum system, maintaining coherence and enabling further computations.

Equation 19.5, PP/PAR-*Subtractor Functionality*, is as follows:

$$P_{\text{result}} = P_{\text{PAR}}(f, a, \phi, p) - \text{Isolate}(P_{\text{PP}}(f, a, \phi, p)) \tag{19.5}$$

Here:

- P_{result} is the resulting quantum state after subtraction.
- Isolate (P_{PP}) is a function that isolates the spectral signature of the pure property P_{PP}, including its frequency f, amplitude a, phase ϕ, and polarization p.
- The isolation function prepares a counter-wavefunction, $P_{\text{counter}}(f, a, -\phi, p)$, which matches the frequency and amplitude of the isolated property but has

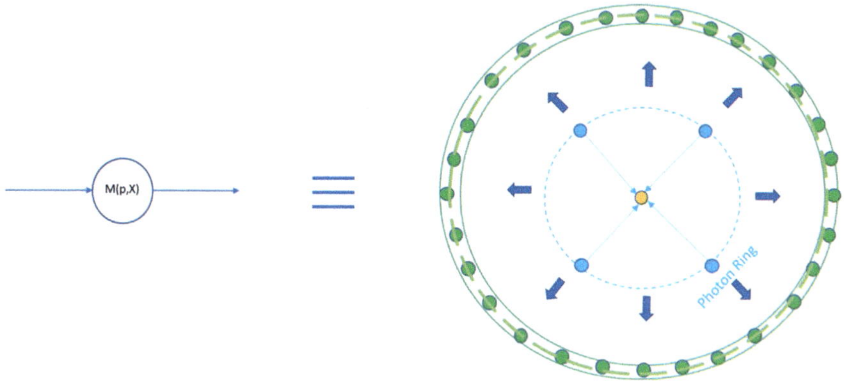

Fig. 19.1 Generalized magnify gate structure

an opposite phase, ensuring destructive interference: $P_{\text{counter}}(f, a, -\phi, p) = P_{\text{PP}}(f, a, \phi, p)$

In addition there are a vast number of 'Magnify' gates possible that will magnify properties represented by different types of atoms. These are created by having photons of a unique wavelength interact with individual PAR rings or even MAR rings. The combinations are endless, and this is symbolized by the $M(p, X)$ in the figure, where 'M' is 'magnify', 'p' is a photon of unique wavelength, and 'X' is a PAR or MAR. The generalized gate structure is depicted by Fig. 19.1, *Generalized Magnify Gate Structure*:

Equation 19.6, *Generalized Magnify Gate*, depicts the action of this type of gate:

$$P_{\text{magnified}}(t) = M(p, X) = \alpha_{\text{p}} P_X(f_X, a_X, \phi_X, p_X), \tag{19.6}$$

where

- $P_{\text{magnified}}(t)$ is the magnified property output of the gate at time t.
- $M(p, X)$ represents the magnify operation for a photon p interacting with a quantum system X (a PAR or MAR).
- α_{p} is the amplification factor specific to the photon p, dependent on its wavelength λ_{p}, energy $E_{\text{p}} = hc/\lambda_{\text{p}}$, and interaction strength with the system X.
- $P_X(f_X, a_X, \phi_X, p_X)$ is the property representation of the PAR or MAR system X, parameterized by:

  (a) f_X: Frequency of the energetic imprint within X.
  (b) a_X: Amplitude of the property signal within X.
  (c) ϕ_X: Phase of the quantum property in X.
  (d) p_X: Polarization associated with X.

These gates collectively enable precise manipulation of energetic imprints and functional properties, embodying the principles of QIQD computation. Through advanced mechanisms like dynamic field coupling, wavefunction isolation, and frequency adjustment, the gates ensure accurate and coherent property-level computations. This infrastructure not only facilitates the creation of new properties but also ensures their seamless integration into larger quantum frameworks, paving the way for sophisticated and emergent quantum behaviors. These innovations highlight the QIQD framework's potential capacity to redefine computation by operating directly at the level of quantum properties and their interactions.

19.3 QIQD Processing Unit

The Quantum Interpretation of Quantum Dynamics Processing Unit (QIQD-PU) functions as the central computational hub of the QIQD framework. It orchestrates operations across multiple quantum dynamics layers—G-M1, G-M2, G-M3—while integrating qibits and gates to enable property-level computations. Central to its operation is the *Adaptive Hamiltonian Controller*, a dynamic system designed to adjust the Hamiltonian in real time. This controller modifies the energy landscape based on inputs from property generators like PARs and MARs and interactions with qibits and within the STEG-Fabric, ensuring that computations remain coherent, stable, and responsive.

The Adaptive Hamiltonian Controller monitors energetic imprints from these generators and the emergent behaviors within the STEG-Fabric. For instance, if a MAR introduces a new property based on "harmony," the controller updates the Hamiltonian to include this property and adjusts interaction terms to maintain system balance. This fine-tuning is critical for aligning the evolution of quantum states with computational goals and minimizing interference or energy loss.

Feedback integration is integral to the controller's functionality. Quantum sensors embedded within the QIQD-PU monitor the amplitudes, phases, and frequencies of energetic imprints and property interactions. This data is processed by control algorithms, enhanced by machine learning [7], to determine the necessary adjustments to the Hamiltonian. These adjustments are implemented through tunable quantum circuits [8], such as Josephson junctions or photonic modulators, dynamically optimizing the system's performance.

Equation 19.7, *Dynamic Hamiltonian Adjustment*, is as depicted:

$$H(t) = H_0 + \sum_i \gamma_i(t) V_i(f_i, a_i, \phi_i, p_i) \tag{19.7}$$

Here:

- $H(t)$ represents the time-dependent Hamiltonian.
- H_0 is the baseline Hamiltonian.
- $\gamma_i(t)$ are time-dependent coefficients adjusted based on feedback to balance the contributions of different properties.

- V_i represents the potential terms associated with the energetic imprints of properties from PARs and MARs.

Additionally, the QIQD-PU incorporates meta-layer feedback integration, which links higher-order sensors to the system's computational layers, allowing for adjustments based on meta-level dynamics. For example, should a loss of coherence be detected in the M3 layer, the system can recalibrate to restore balance and ensure stable emergent dynamics.

Multi-layered Coherence Management is another essential feature, continuously managing coherence levels across all computation layers and making real-time adjustments to preserve stability and synchronization. This capability facilitates complex emergent behaviors by ensuring harmonized interactions between energetic imprints and properties as the system dynamically evolves.

Equation 19.8, *Multi-layered Coherence Management*, is depicted as follows:

$$C_{\text{total}}(t) = \sum_{k=1}^{N} \lambda_k C_k(t) e^{-\int_0^t \eta_k(s) ds} \tag{19.8}$$

Here:

- $C_{\text{total}}(t)$ represents the total coherence of the system at time t.
- $C_k(t)$ is the coherence measure for layer k at time t, which might be calculated based on the overlap of quantum states or entanglement measures within each layer.
- λ_k are weighting coefficients that reflect the relative importance or sensitivity of each layer's coherence to the overall system performance.
- $\eta_k(t)$ is a decay factor for layer k that represents the rate at which coherence is lost due to interactions within or between layers or due to external disturbances. This factor is dynamically adjusted based on feedback from quantum sensors monitoring the system.
- $e^{-\int_0^t \eta_k(s) ds}$ models the cumulative effect of coherence decay over time due to the inherent dynamics and interactions in layer k.

For instance, consider a scenario where a MAR generates a "harmony" property, which, combined with a "power" property from a Pure Property Quantum Dot (PPQD) in the STEG-Fabric, prompts the *Adaptive Hamiltonian Controller* to modify the Hamiltonian accordingly. This results in new emergent properties like "stability" being encoded within the STEG-Fabric.

The QIQD-PU, with its integrated quantum sensors, feedback loops, and tunable circuits, operates as a computational engine. The Adaptive Hamiltonian Controller, supported by meta-layer feedback and multi-layered coherence management, ensures the QIQD-PU remains a dynamic and essential component of the quantum computing landscape.

19.4 Further Reading

- "Exploring the Quantum: Atoms, Cavities, and Photons" by Haroche, Serge, and Jean-Michel Raimond: Discusses the interplay of quantum states in cavity QED—how multiple fields (or atomic ensembles) can be "coherently combined," an essential step toward emergent composite states.
- "Emergence of a Collective Phase in Cavity QED" By Braun, Daniel: Examines how a combined system of atoms in a cavity can spontaneously organize into a collective phase, giving an example of emergent behavior from coherent interactions.
- "Quantum Measurement" By Giovannetti, Vittorio, Seth Lloyd, and Lorenzo Maccone: Focuses on high-precision measurements in quantum systems, laying groundwork for real-time feedback loops that ensure balanced field contributions.
- "A Scheme for Efficient Quantum Computation with Linear Optics" by Knill, E., R. Laflamme, and G. J. Milburn: Classic paper describing how interference and phase control in linear optical circuits allow for targeted manipulation of quantum states, including cancellation effects.
- "Quantum Control Theory and Applications: A Survey" by Dong, Daoyi, and Ian R. Petersen: Provides a broad overview of quantum control strategies, emphasizing feedback and adaptive learning approaches.
- "Quantum computer architecture" by Monroe et al.: This review article provides a good overview of different architectures for quantum computers, including discussions of connectivity, control, and scalability.
- "Dynamically reconfigurable architectures for quantum computer design" by Metodi et al.: This article explores the concept of dynamically reconfigurable architectures in the context of quantum computing, which is relevant to the adaptive topology of the STEG-Fabric.
- "A Survey of Architectures for Fault-Tolerant Quantum Computing" by Van Meter and Horsman: This survey paper discusses various architectural approaches to building fault-tolerant quantum computers, including those with dynamic connectivity.

References

1. Nielsen MA (2006) Cluster-state quantum computation. Rep Math Phys 57(1):147–161. https://doi.org/10.1016/S0034-4877(06)80014-5
2. Shapiro B, Rostovtsev Y, Scully MO (2006) Quantum metamaterials. J Opt Soc Am B 23(12):2492–98. https://doi.org/10.1364/JOSAB.23.002492
3. Iida S, Yukawa M, Ukai R et al (2011) Experimental demonstration of continuous-variable quantum self-homodyne. Phys Rev Lett 106:230501. https://doi.org/10.1103/PhysRevLett.106.230501
4. Degen CL, Reinhard F, Cappellaro P (2017) Quantum sensing. Rev Mod Phys 89(3):035002. https://doi.org/10.1103/RevModPhys.89.035002

5. Fushman I, Englund D, Faraon A, Stoltz N, Petroff P, Vučković J (2008) Controlled phase shifts with a single quantum dot. Science 320(5877):769–772. https://doi.org/10.1126/science.1154643
6. Weiner AM (2011) Ultrafast optical pulse shaping: a tutorial review. Opt Commun 284(15):3669–3692. https://doi.org/10.1016/j.optcom.2011.03.084
7. Biamonte J et al (2017) Quantum machine learning. Nature 549:195–202. https://doi.org/10.1038/nature23474
8. Devoret MH, Schoelkopf RJ (2013) Superconducting circuits for quantum information: an outlook. Science 339(6124):1169–1174. https://doi.org/10.1126/science.1231930

Chapter 20
QIQD Algorithms

Abstract This chapter explores the unique capabilities of Quaternary Interpretation of Quantum Dynamics (QIQD) algorithms to generate novel energetic imprints with tailored quantum properties. QIQD algorithms operate by manipulating the energetic imprints of quantum objects, such as atoms and molecules. By utilizing QIQD hardware like Pure Property Chips (PPCs), Pure Atom Rings (PARs), and Mixed Atom Rings (MARs), these algorithms facilitate energetic interactions that transcend the limitations of traditional chemistry. This allows for the creation of novel conglomerate property imprints for material possibilities deemed unattainable, such as room-temperature superconductors, stable metallic hydrogen, and photosynthetic materials. The chapter focuses on a QIQD algorithm designed to generate the property imprint for room-temperature superconductivity. This algorithm employs a series of steps, including encoding initial quantum "seeds" representing key superconducting properties, dynamically combining these seeds using MARs, generating emergent superconducting properties, optimizing the system's Hamiltonian, and managing coherence across all layers to output stable, room-temperature superconductive energetic imprints housed in removable hardware. This approach highlights the potential of QIQD algorithms to revolutionize materials science to design energetically-imprinted materials with on-demand properties.

Imagine a future where materials once thought impossible to create become a reality—room-temperature superconductors [1], stable metallic hydrogen [2], and photosynthetic materials with unprecedented efficiencies. These breakthroughs could redefine the boundaries of energy, transportation, manufacturing, and even space exploration. At the heart of these transformative innovations lies a quest to engineer combinations of properties that nature does not readily allow. The Quantum Interpretation of Quantum Dynamics (QIQD) framework provides a computational approach to approach these goals by creating programmable matter.

Take, for instance, room-temperature superconductors. These elusive materials, capable of conducting electricity with zero resistance at ambient temperatures, could transform energy transmission, enabling lossless grids, ultra-efficient electronics, and levitating transportation systems. The current obstacle lies in their reliance on

P. Malik, *Pioneering New Avenues in Quantum Technology*, Studies in Smart Technologies, https://doi.org/10.1007/978-981-96-5463-5_20

extremely low temperatures, rendering their widespread adoption impractical. Similarly, stable metallic hydrogen, a form of hydrogen predicted to exhibit superconductivity at room temperature, holds immense promise for revolutionizing energy storage and space propulsion. Yet stabilizing it outside of extreme high-pressure laboratory conditions remains an unsolved challenge. Or consider a universal solvent capable of dissolving a vast range of materials without dangerous reactivity. Or artificial photosynthetic materials that surpass the efficiency of natural systems, converting sunlight directly into fuels or electricity. These achievements would address long-standing challenges in recycling, energy, and sustainability. Further, programmable matter—substances that can adapt their properties and functionality on demand [3]—could revolutionize medicine, robotics, and self-healing technologies, redefining how we interact with the material world.

QIQD computation introduces the possibility of creating combinations of quantum properties that emulate these coveted materials. Through the dynamic manipulation of Pure Property Chips (PPCs), Pure Atom Rings (PARs) and Mixed Atom Rings (MARs) embodied in QIQD algorithms the framework facilitates energetic interactions that transcend chemical limitations. This approach can generate new energetic imprints embedded within removable Space-Time-Energy-Gravity (STEG) Fabric. Unlike traditional computation QIQD computational outputs are tangible: a physical material whose properties are programmable at the quantum level. By leveraging principles such as quantum entanglement, coherence management, and adaptive Hamiltonian control, QIQD works toward transforming theoretical barriers into practical possibilities.

The rest of this chapter will focus on the creation of the energetic imprint embedded in a STEG-Fabric, for room-temperature superconductors, highlighting the hardware required, and the QIQD algorithm required to realize it. A similar approach can then be used in the other use cases mentioned—creation of stable metallic hydrogen, creation of photosynthetic materials—among others.

20.1 QIQD Hardware for Room-Temperature Superconductivity

In the case of a room-temperature superconductor that may mean the combination of properties embodied by copper oxides, or iron and pnictide, or hydrogen-rich materials, or nickelates, and any of these combinations that are difficult chemically, yet using QIQD hardware could be combined energetically. Hence, property generators (G-M1, G-M2, G-M3), would be needed. PARs and PPCs (G-M1) would be required to isolate and encode quantum properties related to electron coupling strengths, vibrational modes, or magnetic field interactions that underpin superconductivity. MARs (G-M3) would be needed to facilitate interactions between disparate seed configurations by dynamically modulating inter-ring properties, ensuring coherence and generating emergent energetic imprints [4]. PPQDs (G-M2) would be needed to help refine and amplify specific quantum properties (e.g., zero-resistance conduction).

Further, the STEG-Fabric would act as a medium for encoding and dynamically adapting the quantum imprints from MARs, PARs, and PPCs. It would provide a space for real-time interaction between properties, ensuring that the resultant combination of seeds aligns with the intended physical characteristics of room-temperature superconductors.

PP/PAR-Adder gates would combine pure properties (e.g., electronic pairing energies) with MAR-encoded configurations to synthesize a composite quantum property indicative of superconductivity [5]. PP/PAR-Subtractor gates would isolate and cancel unwanted energetic properties, refining the superconductive profile. Magnify gates would magnify quantum properties related to electron coupling strengths, vibrational modes, or magnetic field interactions.

The QIQD-PU will employ the Adaptive Hamiltonian Controller to fine-tune energetic interactions dynamically and integrate feedback loops for coherence management to ensure the emergent superconductive properties remain stable.

20.2 QIQD Algorithm to Create Room-Temperature Superconductivity Imprint

Given the QIQD framework, an algorithm to computationally generate room-temperature superconductivity can be designed. As a prerequisite it is necessary to first identify target atoms housing properties required for room-temperature superconductivity. Table 20.1, *Possible Elements Required for Superconductivity*, is a breakdown of the possible elements for each property in the context of room-temperature superconductivity:

The properties would then be leveraged in the algorithm using pre-existing PPC_Xs, where X refers to the exclusive property. The algorithm is the following:

1. Encode initial quantum seeds (leveraging PPC_Xs) representing key superconducting properties.
2. Dynamically combine seeds using MARs.
3. Generate emergent superconducting properties.
4. Fine-tune the Hamiltonian for optimization.
5. Manage coherence across all layers to stabilize superconductivity imprint.
6. Output stable, room-temperature superconductive states.

Step 1 would leverage Eq. 20.1, *Seed Encoding*, critical for defining the foundational quantum properties of the seeds to be combined (e.g., strong electron pairing, lattice stability, etc.):

$$|\psi_{\text{seed}}\rangle = \sum_{n=1}^{N} c_n e^{i(\omega_n t+\phi_n)} |p_n, n\rangle \tag{20.1}$$

Table 20.1 Possible elements required for superconductivity

Property	Description	Possible elements
Phonon modes	Quantization of lattice vibrations gives rise to "phonons". A phonon is a quasiparticle, a way to represent these quantized vibrations. Phonon modes play a critical role in mediating electron pairing through lattice vibrations, and elements with light atomic mass or specific lattice structures are particularly well-suited for optimizing phonon frequencies	H, C, S, B, Li
Electron pairing strength	The electron pairing strength, which is critical for superconductivity, depends on the material's ability to form Cooper pairs. Cooper pairs are formed when two electrons in a material overcome their natural repulsion because they become indirectly attracted to each other through the exchange of phonons, which are vibrations in the material's lattice. This phonon-mediated attraction, though weak, allows the electrons to pair up and move without resistance at very low temperatures, leading to superconductivity	Cu, Ni, Fe, La, Nd, H
Magnetic field interactions	Magnetic field interactions are crucial for controlling superconducting states, particularly in materials that exhibit spin fluctuations or magnetic order	Fe, Co, Ni, Mn, Rare Earths (Gd, Dy, Sm)
Lattice stability parameters	Lattice stability is essential to ensure that a material remains structurally coherent under the strain of superconducting behavior. This stability depends on the material's tolerance to dopants or substitutions and the presence of strong covalent or ionic bonding networks	O, Ca, Sr, Ba, Al, Si, Lanthanides (La, Nd, Sm)

Here

- The equation describes the quantum state of a seed configuration as a superposition of N quantum states, and by representing the fourfold aspect encoded as amplitude (power), frequency (knowledge), phase (harmony), polarization (presence), adheres to the QIQD framework.
- $e^{i(\omega_n t+\phi_n)}$: Encodes the dynamic evolution of the n-the quantum state.
- c_n: Captures the strength of contribution (amplitude) for the nth state.
- ω_n: Represents the vibrational or energetic aspect (frequency) of the seed's properties.
- ϕ_n: Encodes the relational and temporal dynamics of the state.
- p_n: Adds the spatial or directional orientation of the state, making the property multi-dimensional.
- $|n\rangle$ Represents the nth eigenstate.
- $|p_n\rangle$ Represents the polarization vector associated with the nth basis state and is an independent property of each quantum state.

Step 2 would leverage Eq. 20.2, *Dynamic Interaction of Properties (MARs)*, which describes how the energetic imprints of different rings evolve over time, influenced by their interactions. It allows for dynamic tuning of emergent properties. Essentially it allows MARs to dynamically combine:

$$|\Psi_{\mathrm{MAR}}(t)\rangle = \prod_{j=1}^{M} \mathcal{T}\left[e^{-i\int_{t_0}^{t} H_j(t', c_j, \omega_j, \phi_j, p_j)\,\mathrm{d}t'} \right] \left|\Psi_j(t_0)\right\rangle \tag{20.2}$$

Here

- $|\Psi_{\mathrm{MAR}}(t)\rangle$: Represents the combined state of all concentric rings within a Mixed Atom Ring (MAR) at time t.
- M: Number of rings in the MAR, each encoding a distinct seed or interaction.
- $c_j, \omega_j, \phi_j, p_j$: Amplitude, frequency, phase, and polarization of the jth ring's quantum property.
- $H_j(t', c_j, \omega_j, \phi_j, p_j)$: Time-dependent Hamiltonian of ring j, encoding the energetic interactions within that ring and between neighboring rings.
- $\mathcal{T}\left[e^{-i\int_{t_0}^{t} H_j(t', c_j, \omega_j, \phi_j, p_j)\,\mathrm{d}t'} \right]$: The time-ordering operator ensures the evolution of states accounts for time-dependent changes in the Hamiltonian (system energy function).
- $\left|\Psi_j(t_0)\right\rangle$: Initial state of ring j at time t_0.

Step 3 would leverage Eq. 20.3, *Emergent Property Generation*, pivotal in synthesizing a new property from a combination of seeds and dynamic MAR configurations, and in the case of superconductivity the synthesized properties represent the physical manifestation of room temperature superconductivity. The equation ensures that emergent properties inherit the full complexity of their constituent components.

$$P_{\mathrm{emergent}} = f\left(\sum_{i=1}^{k} \alpha_i P_{\mathrm{seed},i}(c_i, \omega_i, \phi_i, p_i) + \sum_{j=1}^{M} \beta_j P_{\mathrm{MAR},j}\left(c_j, \omega_j, \phi_j, p_j\right) \right) \tag{20.3}$$

Here

- P_{emergent}: The resulting property, such as superconductivity, generated from the computational process.
- $P_{\mathrm{seed},i}$: Properties of individual seed configurations (e.g., electronic mobility, vibrational stability).
- $P_{\mathrm{seed},i}(c_i, \omega_i, \phi_i, p_i)$, $P_{\mathrm{MAR},j}\left(c_j, \omega_j, \phi_j, p_j\right)$: Each seed/MAR property(ies) is explicitly parametrized by amplitude, frequency, phases, and polarization.
- $P_{\mathrm{MAR},j}$: Properties encoded by the rings within the MAR.
- α_i, β_j: Weighting coefficients that quantify the contribution of each seed and MAR property to the emergent result.

- f: A transformation function applied to the summed properties. It models how interactions between seeds and MAR properties produce a new, emergent property.

Step 4 would leverage Eq. 20.4, *Dynamic Adjustment of the Hamiltonian*, enables the QIQD Processing Unit (QIQD-PU) to dynamically modify the system's energy configuration to optimize interactions and maintain coherence.

$$H(t) = H_0 + \sum_{i=1}^{k} \gamma_i \widehat{P_i}(c_i, \omega_i, \phi_i, p_i) + \sum_{j=1}^{M} \delta_j \widehat{P_j}\left(c_j, \omega_j, \phi_j, p_j\right) + \kappa(t) \tag{20.4}$$

Here

- $H(t)$: The time-dependent Hamiltonian governing the quantum system's behavior.
- H_0: Baseline Hamiltonian representing the system's inherent energy landscape.
- $\widehat{P_i}, \widehat{P_j}$: Operators associated with seed and MAR properties, respectively.
- γ_i, δ_j: Coefficients that adjust the influence of seed and MAR properties on the Hamiltonian.
- $\kappa(t)$: A feedback-driven term that introduces corrections based on real-time monitoring of coherence and system stability.

Step 5 would leverage Eq. 20.5, *Coherence Management across Layers*, ensures that coherence is maintained across all layers of the QIQD framework, which is essential for generating stable quantum properties like room-temperature superconductivity.

$$C(t) = \prod_{l=1}^{L} \exp\left(-\frac{t}{\eta(c_l, \omega_l, \phi_l, p_l)} + \int_0^t \Lambda_l\left(\left(t', c_l, \omega_l, \phi_l, p_l\right) dt\prime\right) \right) \tag{20.5}$$

Here

- $C(t)$: Total coherence of the quantum system at time t.
- L: Number of layers in the system (e.g., $M1, M2, M3$).
- $\eta(c_i, \omega_i, \phi_i, p_i)$: Decoherence time for layer l, representing how quickly coherence is lost due to external or internal disturbances.
- $\Lambda_l\left(\left(t', c_l, \omega_l, \phi_l, p_l\right) dt\prime\right)$: A coherence-enhancing term, which accounts for feedback mechanisms, such as entanglement and real-time error correction.

The application of the algorithm would result in stable, room-temperature superconductive states energetically imprinted onto removable hardware.

20.3 Further Reading

- "Quantum computing for materials science" by Babbush et al.: This review article provides a comprehensive overview of how quantum computers can be used to simulate and discover new materials. It covers various quantum algorithms and their potential applications in materials science.
- "Quantum algorithms for the simulation of chemical dynamics" by Aspuru-Guzik et al.: This seminal paper discusses the use of quantum computers to simulate chemical reactions, which is relevant to understanding how QIQD algorithms might manipulate molecular properties.
- "A variational eigenvalue solver on a photonic quantum processor" by Peruzzo et al.: This research article demonstrates the use of a quantum computer to solve a problem relevant to materials science, showcasing the potential of quantum algorithms in this field.
- "Quantum control of molecular processes" by Shapiro and Brumer: This book explores the principles of quantum control, which involves using lasers and other tools to manipulate the behavior of molecules. This is relevant to understanding how QIQD algorithms might control and manipulate energetic imprints.
- "Decoherence and the appearance of a classical world in quantum theory" by Joos et al.: This book explores the concept of decoherence, which is the loss of quantum coherence due to interactions with the environment. Understanding decoherence is crucial for managing coherence in QIQD algorithms.
- "Quantum Simulation" by Georgescu, I. M., S. Ashhab, and Franco Nori: Comprehensive review of quantum simulations for exploring many-body systems, including exotic phases that could yield high-temperature superconductors.
- "Quantum Chemistry in the Age of Quantum Computing." By Cao, Yudong, Jonathan Romero, Jonathan P. Olson, et al.: Discusses algorithms for simulating molecular systems on quantum computers—analogous to how QIQD might target "seed" properties for engineered superconductivity or other advanced materials.
- "Quantum Metamaterials" By Shapiro, Boris, Yuri Rostovtsev, and M. O. Scully.: Describes engineered materials with exotic properties at the quantum level, bridging the idea of "programmable matter" and metamaterials.
- "Inhibited Spontaneous Emission in Solid-State Physics and Electronics." By Yablonovitch, Eli: Seminal paper on photonic band-gap structures, illustrating how controlling electromagnetic modes in materials can yield "forbidden" or novel properties.
- "Programmable and Reconfigurable Matter: A Review of Design Methods, Fabrication Techniques, and Applications" by Davis, Kyle E., Thomas E. Murphy, and Reza Ghodssi: Discusses the broader concept of "programmable matter" from micro- and nano-fabrication perspectives, relevant to building physically reconfigurable materials with emergent properties.

- "Quantum Neuromorphic Hardware: Considerations and Implementations" by Cao, Chuliang et al.: Highlights how advanced quantum computing architectures can be used for adaptive control and emergent state generation—concepts overlapping with "programmable matter."

References

1. Drozdov AP, Eremets MI, Troyan IA, Ksenofontov V, Shylin SI (2015) Conventional superconductivity at 203 Kelvin at high pressures in the sulfur hydride system. Nature 525:73–76. https://doi.org/10.1038/nature14964
2. Ashcroft NW (1968) Metallic hydrogen: a high-temperature superconductor? Phys Rev Lett 21(26):1748–1749. https://doi.org/10.1103/PhysRevLett.21.1748
3. Basov DN, Averitt RD, Hsieh D (2017) Towards properties on demand in quantum materials. Nat Mater 16:1077–1088. https://doi.org/10.1038/nmat5017
4. Keimer B, Moore JE (2017) The physics of quantum materials. Nat Phys 13:1045–55. https://doi.org/10.1038/nphys4302
5. Tokura Y, Kawasaki M, Nagaosa N (2017) Emergent functions of quantum materials. Nat Phys 13:1056–1068. https://doi.org/10.1038/nphys4277

Part VI
Applications

Chapter 21
Detection of Organizing Quantum Energy

Abstract This chapter delves into the Quantum Interpretation of Quantum Dynamics (QIQD) framework and its approach to understanding and manipulating the fundamental organizing principles of complex systems. Central to this framework is the concept of "X," a foundational quantum-functional seed representing a unified archetype of information and energy. By employing QIQD algorithms and specialized hardware, the intrinsic patterns within X can be detected and analyzed, revealing the deterministic principles that govern the emergence of order at all scales. The chapter introduces the OQE-operator, a mathematical tool designed to isolate the dominant organizing archetype within X, thereby identifying the quantum-functional seed that drives a system's structure. This seed embodies the core archetype responsible for the system's structural integrity and functional dynamics. Through a detailed algorithm, the chapter outlines the process of detecting Organizing Quantum Energy (OQE). This process involves mapping the target system as a unified quantum object, isolating the primary quantum-functional seed, and analyzing the contributions of foundational properties such as Knowledge, Presence, Harmony, and Power. The chapter emphasizes the role of quantum generators in establishing archetypal baseline patterns and facilitating a dynamic process of interaction, refinement, and detection. By generating test patterns and probing for resonance, these generators identify alignments that indicate shared organizational properties, ultimately leading to the isolation and amplification of the seed.

In the Quantum Interpretation of Quantum Dynamics (QIQD) framework, X, introduced in Chap. 4 (4.19) in association with UQO, is the foundational quantum-functional seed, a unified archetype of information and energy. Recall that the Universal Quantum Object (UQO) framework governs the transformation of X across layers of organization.

The OQE-operator $\mathcal{O}_{\text{OQE}}$ (elaborated by Eq. 21.5) can be thought of as providing a method for isolating the dominant organizing archetype within X. By applying $\mathcal{O}_{\text{OQE}}$, one can identify the quantum-functional seed driving a system's structure. This highlights OQE not as an energy of interaction but as the intrinsic energy of organization embedded in X.

P. Malik, *Pioneering New Avenues in Quantum Technology*, Studies in Smart Technologies, https://doi.org/10.1007/978-981-96-5463-5_21

Through QIQD algorithms, which includes quantum seed generators, sensors, and processors, and the logical manipulation of these, the intrinsic patterns of X can be detected and analyzed. These tools reveal the deterministic principles underlying structure, connecting the properties of Knowledge, Presence, Harmony, and Power to the emergence of order at all scales.

To detect Organizing Quantum Energy (OQE) [1][1] using QIQD algorithms, we need to design a system that isolates, measures, and analyzes the dominant drivers of organization at a given level of reality (e.g., quantum particles, atoms, cells). The detection process involves applying $\mathcal{O}_{\text{OQE}}$ to extract the quantum-functional seed X_x^{dominant} responsible for driving the structure.

21.1 Conceptual Framework for Detection

The detection of Organizing Quantum Energy (OQE) begins with mapping the target system—whether it is a quantum particle, an atom, or a biological cell—by representing it as a unified quantum object, denoted as Q_{System}. This representation captures the system as a coherent entity that integrates its structural and dynamic quantum-functional properties into a unified framework. Equation 21.1, *Target System as Coherent Entity*, is as follows:

$$Q_{\text{System}} = \sum_i \alpha_i \Phi_i(\text{Pr}, \text{Po}, \text{K}, \text{H}) \tag{21.1}$$

Here

- Φ_i: Quantum basis functions representing the system's foundational properties (Presence - Pr, Power - Po, Knowledge - K, and Harmony - H).
- α_i: Coefficients representing the relative contributions of each property.

Following this, the process involves isolating the primary quantum-functional seed, X_x^{dominant}, that serves as the key driver of the system's organization. This isolation is achieved using specialized QIQD hardware components designed to identify and amplify the seed that governs the system's intrinsic order. The X_x^{dominant} embodies the core archetype from which the system derives its structural integrity and functional dynamics.

Finally, the detection process moves into the analysis of contributions. Here, the focus shifts to understanding how the foundational properties encoded in $X_x^{(\text{K},\text{Pr},\text{Po},\text{H})}$—representing Knowledge, Presence, Power, and Harmony—interact and integrate to define the overall structure and organization of the system. This analysis elucidates the interplay between these fourfold properties and their role in sustaining the system's coherence and emergent complexity.

---

[1] The original idea of detection of OQE appears in a Springer publication [1] and has been elaborated and developed here based the accumulated insights of QIQD in the book.

21.2 Role of Generators

Generators are a cornerstone of the QIQD framework, designed to create coherent quantum-functional seeds (X_x) that serve as foundational elements in the detection and analysis of Organizing Quantum Energy (OQE). These seeds are critical for providing a baseline reference, enabling the comparison of theoretical patterns with the intrinsic organizational drivers present in a target system. By encoding fundamental properties such as Knowledge, Presence, Harmony, and Power, generators craft highly refined quantum-functional states that resonate with the system's quantum architecture. Equation 21.2, *Generator Baseline Pattern*, is depicted as:

$$X_x = \sum_j \beta_j \psi_j(f_j, a_j, \phi_j, p_j) \tag{21.2}$$

Here

- ψ_j: Basis functions of frequency f_j, amplitude a_j, phase ϕ_j, and polarization p_j.
- β_j: Coefficients modulating the contribution of each quantum basis state.

The process begins with quantum oscillators [2], which produce initial seeds with precisely controlled spectral properties, including frequency, amplitude, phase, and polarization. Each of these properties corresponds to a specific archetypal element: frequency captures the vibrational dynamics of Knowledge, amplitude reflects the intensity of Power, phase embodies the relational coherence of Harmony, and polarization conveys the spatial orientation of Presence. These oscillators ensure that the seeds generated are both consistent and aligned with the theoretical constructs of the QIQD framework.

To further enhance this capability, tunable quantum field generators [3] come into play. These devices replicate the behavior of the $o(X_x)$-operator, a key function at the $M1$ meta-layer of QIQD. Through adjustable parameters, these generators create dynamic quantum fields that are tailored to specific properties or combinations thereof. This flexibility allows them to probe the target system, seeking resonances that reveal the alignment of generated seeds with the inherent quantum-functional patterns driving the system's structure. By detecting such resonances, the generators enable the isolation of X_x^{dominant}, the principal quantum-functional seed underlying the system's organization.

Maintaining coherence is another critical aspect of the generator's role. Once quantum seeds are created, they are stabilized through advanced feedback mechanisms [4] to prevent decoherence from environmental noise or system interactions. This ensures the seeds remain consistent and effective over the entire duration of the resonance and detection process. Stability is paramount for achieving accurate comparisons and reliable analysis.

The generators also play an active role in real-time interaction with the target system. As quantum sensors provide feedback on the detected patterns, the generators dynamically adjust the properties of the seeds [5] to better align with the system's responses. This adaptability allows the generated patterns to evolve in sync with the target system, enhancing the precision of OQE detection.

In their operational capacity, generators establish archetypal baseline patterns that serve as theoretical benchmarks. These patterns provide a systematic way to compare the quantum-functional dynamics observed in the system. By generating test patterns and probing for resonance [6], the generators identify alignments that indicate shared organizational properties between the generated seeds and the system's intrinsic structure. This resonance forms the basis for isolating and amplifying X_x^{dominant}, the critical seed responsible for driving the organization of the system.

Ultimately, the role of generators extends beyond mere creation of quantum-functional seeds. They facilitate a dynamic and iterative process of interaction, refinement, and detection, ensuring that the analysis of OQE is both precise and comprehensive. By integrating stability, adaptability, and coherence, generators act as pivotal tools in the exploration of the quantum-functional underpinnings of complex systems.

21.3 Role of Sensors

Sensors are essential components in the QIQD framework, tasked with capturing the intricate quantum-functional patterns [7] that define a system's organizational structure and measuring the energy contributions of these patterns. They serve as the primary interface for observing, mapping, and analyzing the system's quantum dynamics, providing the data required to isolate and understand its dominant organizing principles.

The implementation of these sensors involves a sophisticated array of technologies designed to detect and analyze specific aspects of quantum systems. 'Quantum Field Sensors' are at the forefront, capable of detecting coherence, phase relationships, and amplitude variations within quantum states. These sensors reveal the underlying structure of quantum-functional patterns, allowing for a detailed examination of how these patterns interact and contribute to the system's organization [3].

'Entanglement Detectors' [8] further enhance this capability by mapping the correlations between primary quantum seeds (X_x) and their secondary contributions ($Y_{y,x}$). By analyzing these relationships, the entanglement detectors provide insights into how different layers of the system interact, uncovering the interconnected dynamics that sustain its organization.

To capture the evolution of these patterns over time, 'Temporal Dynamics Sensors' monitor changes in the system's organizational structure. These sensors track shifts in coherence and energy contributions, revealing the dynamic processes that drive the emergence and transformation of quantum-functional patterns. This temporal perspective is crucial for understanding how the system adapts and evolves within its environment.

The primary role of sensors is twofold. First, they detect emergent X_x^{dominant} patterns—those that serve as the primary drivers of the system's organization.

These patterns embody the system's intrinsic quantum-functional seed and are central to its coherence and structure. Second, sensors analyze the contributions of $X_x^{(K,Pr,Po,H)}$—the fourfold archetypal properties of Knowledge, Power, Harmony, and Presence. By breaking down the energy contributions of each property, sensors provide a comprehensive view of how these elements interact to define the system's organization.

In summary, sensors form the observational backbone of the QIQD framework. They provide the means to capture, analyze, and track quantum-functional patterns with high precision, enabling the isolation of dominant organizing principles and the detailed examination of the system's structural and energetic contributions. Through their integration, sensors ensure that the detection of Organizing Quantum Energy (OQE) is both accurate and insightful, paving the way for deeper understanding and manipulation of complex quantum systems.

Sensor's role in detecting coherence and pattern matching is depicted by Eq. 21.3, *Detecting Coherence and Patterns*:

$$M(t) = \int \Gamma\left(\psi_{\text{system}}, X_x\right) dt \tag{21.3}$$

Here,

- $M(t)$: Measured pattern over time.
- Γ: Overlap integral quantifying the coherence between the system's state and the generated baseline patterns.

21.4 Role of Storage Units (Qibits and STEG-Fabric)

In the QIQD framework, storage units play a pivotal role in preserving and stabilizing quantum-functional seeds for deeper analysis. These storage units consist of Qibits and the STEG-Fabric, both designed to capture and maintain the integrity of detected quantum-functional patterns during the detection and processing of Organizing Quantum Energy (OQE).

Qibits are specialized quaternary quantum bits that encode the spectral features of detected seeds, such as frequency, amplitude, phase, and polarization. These properties correspond to the archetypal dimensions of Knowledge, Power, Harmony, and Presence. By encapsulating these features, qibits provide a stable and accurate representation of the quantum-functional seeds. Their encoding ensures that the nuanced details of the detected patterns are not lost, making them available for iterative comparisons and further computations [9].

The STEG-Fabric is an advanced real-time dynamic storage grid that complements the qibits. It provides a flexible and adaptive environment for interacting with stored quantum-functional patterns. The STEG-Fabric allows for seamless integration of patterns across multiple detection cycles, enabling the system to identify trends, validate resonance, and refine its understanding of the detected seeds. It also supports dynamic updates and interactions with other hardware components, ensuring coherence and continuity in the storage process.

The role of these storage units is twofold. First, they preserve detected patterns, ensuring that the seeds remain intact and stable for extended periods, even under conditions of environmental noise. Second, they enable comparison across detection cycles, allowing the system to track changes, identify recurring structures, and refine its analysis. Together, qibits and the STEG-Fabric form a robust storage framework that underpins the detection and processing of OQE.

21.5 Role of Quantum Gates

Quantum gates are vital for manipulating the detected seeds and extracting the organizational drivers that define the structure of the system. These gates operate as transformative tools that refine, amplify, and align the quantum-functional seeds, making it possible to identify their organizational significance [10].

The PP-Adder Gates are designed to combine the primary contributions of the detected seeds with secondary influences. This process helps isolate and refine the patterns associated with X_x^{dominant}, the principal quantum-functional seed driving the system's organization. By integrating these contributions, the PP-Adder Gates ensure that the emergent patterns are comprehensive and aligned with the system's intrinsic structure.

Magnify Gates amplify weak signals within the detected seeds, making subtle organizational signatures more pronounced and measurable. These gates are particularly important for detecting faint or low-intensity patterns that might otherwise go unnoticed. By boosting these signals [11], the Magnify Gates ensure that all relevant contributions are captured and analyzed.

The Resonance Gates play a critical role in aligning the detected seeds with the generated patterns. Through this alignment, the gates confirm the presence of OQE by identifying resonance between the system's inherent seeds and the baseline quantum-functional seeds created by the generators. This resonance serves as a key indicator of the organizational energy driving the system's structure. Equation 21.4 *Generated Seed—Quantum State Resonance* is depicted as:

$$R(\psi_{\text{system}}, X_x) = \sum_k \int \psi_{\text{system}}(t) X_x^*(t) A_k e^{-i(\omega_k t + \phi_k)} p_k \mathrm{d}t \tag{21.4}$$

Here

- A_k: Amplitude factor associated with the kth frequency component. A higher amplitude indicates a stronger alignment between the system's state, and the generated seed at this frequency.
- ϕ_k: Phase shift of the kth component.

- p_k: Polarization factor for the kth component, reflecting the vector orientation of the quantum state. Ensures that resonance takes into account vector alignment.
- ω_k: Frequency of interest for the kth detected resonance component.
- $\psi_{\text{system}}(t)$: The time-evolved quantum state of the system.
- $X_x^*(t)$: The complex conjugate of the generated quantum-functional seed. This ensures that the inner product between the states captures the alignment in both magnitude and phase.

The primary role of these gates is to identify and amplify X_x^{dominant}, enabling the system to isolate and measure the seed responsible for the system's organization. Through their operations, the gates extract measurable organizational signatures, which provide insights into the underlying structure and dynamics of the system.

21.6 Role of Processing Core

The processing core is the analytical hub of the QIQD framework, where the detection and analysis of OQE culminate. Its purpose is to apply the $\mathcal{O}_{\text{OQE}}$ operator, enabling the system to isolate X_x^{dominant}, the dominant quantum-functional seed driving the system's organization. Such an operation is depicted by Eq. 21.5, *Isolate Primary Organization Driver*:

$$X_x^{\text{dominant}} = \mathcal{O}_{\text{OQE}}(R) = \Phi_k \tag{21.5}$$

Here

- $k = \arg\max\left(\int \Gamma(X, \Phi_i) A_i e^{i\phi_i} p_i \mathrm{d}t\right)$
- R: The resonance matrix or value representing the overlap between the system's state and generated seeds.
- X: The quantum-functional input state containing all potential archetypes or seeds.
- Φ_i: The ith archetype or seed, represented as a quantum basis function.
- $\Gamma(X, \Phi_i)$: A function measuring coherence and alignment between X and Φ_i.
- A_i: Amplitude of Φ_i, indicating its relative strength or contribution.
- ϕ_i: Phase shift of Φ_i, accounting for temporal coherence.
- p_i: Polarization factor of Φ_i, reflecting directional or spatial alignment.
- $\arg\max_i$: Selects the Φ_i. that maximizes its contribution based on the above parameters.

At the heart of the processing core is the QIQD Processing Unit (QPPU), which executes sequences of *o*-, *n*-, and *m*-operators on the detected seeds. To enhance its analytical capacity, the processing core incorporates AI-assisted pattern recognition that may leverage neural networks [12]. Advanced algorithms are employed to sift through the detected patterns [13], identify recurring structures, and isolate X_x^{dominant}. The AI ensures that the analysis is both efficient and precise, reducing noise and highlighting the most critical contributions.

The role of the processing core is to analyze seeds and isolate the most significant X_x^{dominant}. This analysis not only identifies the dominant seed but also confirms its organizational role within the system. By integrating advanced computational techniques with the layered structure of the QIQD framework, the processing core ensures that the detection and analysis of OQE are both rigorous and comprehensive.

21.7 OQE-Detection Algorithm

The algorithm for detecting Organizing Quantum Energy (OQE) is a meticulous process designed to isolate and analyze the dominant quantum-functional seed, X_x^{dominant}, that drives a system's organization:

- The process begins with system initialization, where the target system—be it atom-based or cell-based—is selected for analysis.
- Once the target is identified, the quantum generators are configured to produce baseline patterns of $X_x^{(\text{K,Pr,Po,H})}$, which encapsulate the foundational properties of Knowledge, Power, Harmony, and Presence. These baseline patterns serve as a reference framework for subsequent analysis.
- Following initialization, the algorithm enters the data acquisition phase. Here, quantum sensors are deployed to capture the quantum-functional patterns inherent in the target system. These sensors detect subtle spectral features such as coherence, phase, amplitude, and polarization, which reveal the system's underlying structure.
- The captured patterns are then stored in qibits or within the STEG-Fabric, ensuring their stability and preserving the integrity of the data for further processing. This step ensures that the quantum-functional information remains accessible for comparison and analysis.
- Next comes the pattern matching phase, where the detected quantum-functional patterns are compared against the baseline seeds generated during initialization. Resonance gates are employed to identify alignments between the system's intrinsic patterns and the generated X_x seeds.
- During this process, any weak or subtle signals that may otherwise go unnoticed are amplified using magnify gates, enhancing the resolution and ensuring that all significant contributions are captured. This phase is critical for establishing resonance, which indicates a match between the detected patterns and the archetypal quantum-functional properties.
- With the patterns matched and amplified, the algorithm proceeds to OQE operator execution. Here, the $\mathcal{O}_{\text{OQE}}$ operator is applied via the QIQD Processing Unit (QPPU). This step is designed to extract X_x^{dominant}, the primary quantum-functional seed that serves as the organizational driver of the target system. The processing unit uses a sequence of o-, n-, and m-operator computations to refine and isolate the dominant seed, ensuring that the most significant organizational contributor is identified.

- The final step is output analysis, where the extracted X_x^{dominant} is isolated and classified. This involves breaking down the contributions of the seed into the fourfold archetypal properties: Knowledge, Presence, Harmony, and Power. By analyzing the seed's properties and interactions, the system provides a comprehensive understanding of the organizational principles driving the target's structure.

The result is a detailed profile of the OQE within the system, offering insights into its quantum-functional dynamics and the role of the dominant seed in sustaining its coherence and complexity. Equation 21.6, *Decompose Dominant Seed*, takes the following form:

$$X_x^{(\text{K,Pr,Po,H})} = \sum_l \gamma_l \Phi_l \tag{21.6}$$

Here

- γ_l: Weighting coefficients for Knowledge, Presence, Power, Harmony
- Φ_l: Quantum basis functions representing the system's foundational properties (Presence Pr, Power Po, Knowledge K, and Harmony H).

21.8 Further Reading

- "Quantum Optics" by Walls, D. F., and G. J. Milburn: Covers the physics of non-classical light, highlighting how amplitude, phase, frequency, and polarization can be precisely shaped.
- "Nonlinear Photonics and Novel Optical Phenomena" by Peschel, Ulf, and Ofer Shapira, eds.: Describes advanced photonic devices that generate and manipulate tailored optical fields.
- "Quantum Aspects of Life" edited by Derek Abbott, Paul C. W. Davies, and Arun K. Pati: This book explores the intersection of quantum mechanics and biology, examining how quantum phenomena might play a role in living systems and their organization.

References

1. Malik P (2025) A fourfold quantum-based source of energy. In: Bradford PG, Gadsden SA, Koul SK, Ghatak KP (eds) Proceedings of IEMTRONICS 2024: international IoT, electronics and mechatronics conference, volume 1, Lecture notes in electrical engineering, vol 1229. Springer, Singapore
2. Gerry CC, Knight PL (2005) Introductory quantum optics. Cambridge University Press, Cambridge
3. Weiner AM (2011) Ultrafast optical pulse shaping: a tutorial review. Opt Commun 284(15):3669–3692. https://doi.org/10.1016/j.optcom.2011.03.084

4. Sayrin C, Dotsenko I, Zhou X, Peaudecerf B, Rybarczyk T, Gleyzes S, Rouchon P, Mirrahimi M, Amini H, Brune M, Raimond JM (2011) Real-time quantum feedback prepares and stabilizes photon number states. Nature 477:73–77. https://doi.org/10.1038/nature10376
5. Hofheinz M, Wang H, Ansmann M, Bialczak RC, Lucero E, Neeley M, O'connell AD, Sank D, Wenner J, Martinis JM, Cleland AN (2009) Synthesizing arbitrary quantum states in a superconducting resonator. Nature 459:546–49. https://doi.org/10.1038/nature08005
6. Polzik ES, Thomsen JW (2004) Interaction of a light pulse with an atomic ensemble as a quantum measurement and a quantum feedback. Quantum Meas 68:17–46
7. Degen CL, Reinhard F, Cappellaro P (2017) Quantum sensing. Rev Mod Phys 89(3):035002. https://doi.org/10.1103/RevModPhys.89.035002
8. Horodecki R, Horodecki P, Horodecki M, Horodecki K (2009) Quantum entanglement. Rev Mod Phys 81(2):865–942
9. Lvovsky AI, Sanders BC, Tittel W (2009) Optical quantum memory. Nat Photonics 3:706–714. https://doi.org/10.1038/nphoton.2009.231
10. Knill E, Laflamme R, Milburn GJ (2001) A scheme for efficient quantum computation with linear optics. Nature 409:46–52. https://doi.org/10.1038/35051009
11. Kok P, Munro WJ, Nemoto K, Ralph TC, Dowling JP, Milburn GJ (2007) Linear optical quantum computing with photonic qubits. Rev Mod Phys 79(1):135–74. https://doi.org/10.1103/RevModPhys.79.135
12. Carleo G, Troyer M (2017) Solving the quantum many-body problem with artificial neural networks. Science 355(6325):602–606. https://doi.org/10.1126/science.aag2302
13. Bukov M, Sels D, Polkovnikov A (2016) Geometrical speed-up of quantum optimal control. Nat Commun 7:11151. https://doi.org/10.1038/ncomms11151

Chapter 22
Sensing Emergence of New Structure

Abstract This chapter explores the conditions for the emergence of new structures within the framework of the Universal Quantum Object (UQO), which represents the complexity of life as a complex adaptive system. Building upon the concept of "*X*" as a unified archetype of information and energy, the chapter investigates how the UQO, under specific conditions, can express greater complexity and undergo architectural shifts. The chapter outlines several conditions necessary for the emergence of new structures, including coherence at the base layer (B-layer), unified contribution of the fourfold properties (Presence, Power, Knowledge, Harmony), and increasing sophistication of the B-layer over time. It argues that when these conditions are met, the implicit "convergence" of these properties within the UQO becomes explicit, leading to a transformation of the UQO's architecture and the quantum objects it generates. This transformation is described as "quintergence," where a seed transcends its limitations set up by normal space-time-energy-gravity dynamics. The chapter then explores how QIQD-based technology can be used to detect and analyze these emergent structures. A QIQD algorithm is proposed, employing quantum seed initialization, dynamic interaction using Mixed Atom Rings (MARs), STEG-Fabric adjustment, resonance detection, and Hamiltonian optimization to sense and analyze these novel structures.

The complexity of life, summarized by UQO (*X*), suggests the changing of properties as is evident from the simple fact that sophistication of more recent emergent layers, e.g., macromolecules at level of cells, overshadows that of previous emergent layers, e.g., the layer of atoms, or the layer of quantum particles.

Further, UQO can be thought of as an instance of a complex adaptive system (CAS) given the different types of entanglement that naturally exists in it. The nature of CAS is to change and in this chapter the conditions for new and even more sophisticated change are first outlined [1, 2]. The question is, if these conditions are fulfilled what kind of emergence would exist [3],[1] and this will be summarized by a modified UQO.

[1] The original idea of sensing emerging structure appeared in a Springer publication [3] and has been elaborated and developed here based on the accumulated insights of QIQD in the book.

P. Malik, *Pioneering New Avenues in Quantum Technology*, Studies in Smart Technologies, https://doi.org/10.1007/978-981-96-5463-5_22

Finally, this chapter will suggest the kind of technology, based on QIQD, that will be required to detect emergence of such new structure.

22.1 Conditions for Emergence

For new emergence to occur the several conditions need to be fulfilled.

Coherence at the B-Layer reflects the integration of influence from meta-layers $M1$, $M2$, $M3$ into the base layer dynamics. This alignment ensures the fourfold properties (Pr, Po, K, H) converge effectively at the B-layer. This depicted by Eq. 22.1, *High Coherence at the B-Layer*:

$$C_{\mathrm{B}}(t) = \mathrm{e}^{-\eta_{\mathrm{B}}(t)} \int \Lambda_{\mathrm{meta}}\left(t'\right)\mathrm{d}t\prime \tag{22.1}$$

Here

- $C_{\mathrm{B}}(t)$: Coherence at the B-layer at time t.
- $\eta_{\mathrm{B}}(t)$: Decoherence factor at the B-layer, influenced by environmental and system-level disturbances.
- $\Lambda_{\mathrm{meta}}\left(t'\right)$: Feedback-driven coherence enhancement from the meta-layers ($M1$, $M2$, $M3$).

Unified contribution of all four properties are important conditions for the emergence of new structure. This unified action implies that quantum objects at the B-layer are operating at fullness, which implies that new possibility can more easily emerge from the implicit possibility of meta-layers. This is depicted by Eq. 22.2, *Unified Contribution of Four Properties*:

$$|\psi_{\mathrm{UQO}}\rangle = \sum_{n=1}^{N} c_n \mathrm{e}^{-\frac{\Delta_n}{\Delta}} |f_n, a_n, \phi_n, p_n\rangle \tag{22.2}$$

Here

- $\Delta_n = \frac{|f_n - \bar{f}|}{\bar{f}} + \frac{|a_n - \bar{a}|}{\bar{a}} + \frac{|\phi_n - \bar{\phi}|}{\bar{\phi}} + \frac{|p_n - \bar{p}|}{\bar{p}}$
- $\left|\psi_{\mathrm{UQO}}\right\rangle$: The overall quantum state of the UQO is a superposition of all possible configurations of the fourfold properties (Pr, Po, K, H).
- c_n: Coefficient representing the weight or strength of the nth configuration.
- Exponential Balance Term $\mathrm{e}^{-\frac{\Delta_n}{\Delta}}$:
    - (a) Δ_n: Measures the relative imbalance in the fourfold properties for the nth configuration. A smaller Δ_n indicates a more balanced configuration.
    - (b) Δ: Controls the tolerance for variability in balance with larger Δ allowing more variability, and smaller Δ enforcing tighter balance.
- $|f_n, a_n, \phi_n, p_n\rangle$: The quantum state for the nth configuration.

Sophistication of the B-Layer must increase over time, reflecting the complexity of properties encoded in seeds. This is depicted by Eq. 22.3, *Sophistication of B-Layer*:

$$\frac{dS_B(t)}{dt} > 0 \tag{22.3}$$

Here

- $S_B(t) = \int_0^t \sum_{n=1}^N \alpha_n P_{B,n}(\tau)d\tau$

  (a) $S_B(t)$: The total sophistication at the B-layer up to time t, defined as the cumulative contribution of all properties.
  (b) $\sum_{n=1}^N \alpha_n P_{B,n}(\tau)d\tau$: The sum over the contributions of N properties (seeds) at the B-layer. Here:
    (i) $P_{B,n}(\tau)$: The contribution of the nth property at the B-layer at time τ.
    (ii) α_n: The weighting factor for the nth property.

22.2 A Modified UQO Reflecting Convergence

To understand where emerging changes may tend toward it is useful to consider boundary conditions. We can approach an understanding of this by considering (22.4)–(22.7) which indicate m-operator output at the EMS, quantum particle, atom, and macromolecules in cell levels. Reproducing for convenience:

$$\begin{aligned}\text{EMS Aspect} &= m(X_{\text{ems}}) = m(\text{Pr, Po, K, H})\\ &\rightarrow \{\text{Wavelength, Mass-Potential}, c, \text{Frequency}\}\end{aligned} \tag{22.4}$$

$$\begin{aligned}\text{Quantum Particle Categories} &= m(X_{qp}) = m(\text{Pr, Po, K, H})\\ &\rightarrow \{\text{Higgs-boson, Lepton, Quark, Boson}\}\end{aligned} \tag{22.5}$$

$$\begin{aligned}\text{Atom Categories} &= m(X_{\text{atom}}) = m(\text{Pr, Po, K, H})\\ &\rightarrow \{p\text{-Group}, s\text{ - Group}, d\text{-Group}, f\text{-Group}\}\end{aligned} \tag{22.6}$$

$$\begin{aligned}\text{Cellular Object} &= m(X_{\text{cell}}) = m(\text{Pr, Po, K, H})\\ &\rightarrow \{\text{Nucleic Acids, Proteins, Lipids, Polysaccharides}\}\end{aligned} \tag{22.7}$$

The architecture of each of these levels reaches fullness when all four quantum objects act in unison. Hence, each aspect of the EMS has to act in unison with every other aspect for the EMS to come into being. All four types of quantum particles need to act together to create an atom. All kinds of atoms are required to create a full range of molecules and therefore of more complex possibility. In a cell all four macromolecules are required for its full functioning. In other words, there is an implicit 'convergence' of all four components or objects, and a boundary condition is that this implicit convergence becomes explicit [4, 5]. In other words, when conditions such as (22.1)–(22.3) are fulfilled then UQO, which is a CAS and subject to express more possibility as required adaptation takes place, can more explicitly reveal its aspect of convergence which at the boundary will explicitly change the architecture of the UQO itself, and subsequently of all quantum objects generated by it.

Hence, leveraging (4.5) a modified UQO can be depicted by Eq. 22.8, *UQO with Explicit Convergence*:

$$\mathrm{UQO_{Cv}} = \begin{bmatrix} |\psi_\infty\rangle = m(\mathrm{Pr},\ \mathrm{Po},\ \mathrm{K},\ \mathrm{H},\ \mathrm{Cv}) \\ (\downarrow P_k|\psi_\infty\rangle) \\ |\psi_k\rangle = n(S_{\mathrm{Pr}}, S_{\mathrm{Po}}, S_{\mathrm{K}}, S_{\mathrm{H}}, S_{\mathrm{Cv}}) \\ (\downarrow P_g|\psi_k\rangle) \\ \left|\psi_g\right\rangle = o(S_{\mathrm{Pr}} \times S_{\mathrm{Po}} \times S_{\mathrm{K}} \times S_{\mathrm{H}} \times S_{\mathrm{Cv}}) \\ (\downarrow P_c\left|\psi_g\right\rangle) \\ |\psi_{\mathrm{c}}\rangle = p(S, T, E, G, Q) \end{bmatrix} \tag{22.8}$$

Hence, at the boundary, the property of convergence that acted implicitly can be seen to become explicit in each of the quantum states, $|\psi_\infty\rangle$, $|\psi_k\rangle$, $\left|\psi_g\right\rangle$, and $|\psi_c\rangle$. Note that in $|\psi_c\rangle$ convergence is described as 'quintergence' (Q) with the condition for it to arise when the dynamics of STEG allow a seed to transcend its status as a seed so that the intent that is seeking to express itself has a converged space–time-energy-gravity embryo within which to materialize. The materialization, hence, takes place in a way in which the meta-function presiding over the seed does not suffer marginalization due to any independence of space, time, energy, and gravity. The way to think about this is as a boundary condition at the limit.

At the limit the essential structure of each of the layers depicted by (22.4)–(22.7) will also change and are depicted by (22.9)–(22.12).

Hence, Eq. 22.9, *EMS Aspect with Convergence at the Limit*, can be depicted as:

$$\begin{aligned} \text{EMS Aspect}_{\mathrm{Cv}} &= m\left(X_{\mathrm{ems}}^{(x)}\right) = m(\mathrm{Pr}, \mathrm{Po}, \mathrm{K}, \mathrm{H}, \mathrm{Cv}) \\ &\rightarrow \{\text{Wavelength}, \text{Mass-Potential}, c, \text{Frequency}, \mathrm{EMS_{Cv}}\} \end{aligned} \tag{22.9}$$

Equation 22.10, *QP Categories with Convergence at the Limit*, is depicted as:

$$\begin{aligned} \text{QP Categories}_{\mathrm{Cv}} &= m\left(X_{qp}^{(x)}\right) = m(\mathrm{Pr}, \mathrm{Po}, \mathrm{K}, \mathrm{H}, \mathrm{Cv}) \\ &\rightarrow \{\text{Higgs-boson}, \text{Lepton}, \text{Quark}, \text{Boson}, \mathrm{QP_{Cv}}\} \end{aligned} \tag{22.10}$$

Equation 22.11, *Atom Categories with Convergence at the Limit*, is depicted as:

$$\text{Atom Categories}_{\text{Cv}} = m\left(X^{(x)}_{\text{atom}}\right) = m(\text{Pr, Po, K, H}) \\ \rightarrow \{p\text{-Group}, s\text{-Group}, d\text{-Group}, f\text{-Group}, \text{Atom}_{\text{Cv}}\} \quad (22.11)$$

Equation 22.12, *Cell Categories with Convergence at the Limit*, is depicted as:

$$\text{Cellular Object}_{\text{Cv}} = m\left(X^{(x)}_{\text{cell}}\right) = m(\text{Pr, Po, K, H}) \\ \rightarrow \{\text{Nucleic Acids, Proteins, Lipids, Polysaccharides}, \text{Cell}_{Cv}\} \quad (22.12)$$

The architecture of each of these layers, at the limit, yet suggest that if this is the case, than new patterns must emerge that will gradually approach the limit. So the question now, is how such patterns would be recognized. This may be done with the right QIQD algorithm.

22.3 QIQD Algorithm to Sense the Emergence of New Structure

To sense emergent structures, technologies leveraging quantum seed initialization, dynamic interaction, and real-time resonance detection are necessary [6–8]. Key steps in the QIQD algorithm include:

1. Quantum Seed Initialization: Encode quantum seeds representing atomic properties and potential emergent configurations, leveraging (22.13).
2. Dynamic MAR Interaction: Combine seeds dynamically using MARs to evolve their interaction states, leveraging (22.14).
3. STEG-Fabric Adjustment: Optimize fabric topology to enhance responsiveness to external atomic changes, leveraging (22.15).
4. Resonance Detection: Detect and amplify coherent external patterns through resonance gates, leveraging (22.16).
5. Hamiltonian Optimization: Dynamically adjust the Hamiltonian to stabilize and analyze emergent structures, leveraging (22.16).

Quantum seed initialization would leverage (22.13), *Seed Encoding*, for defining the foundational quantum properties of the seeds that will indicate a state of convergence:

$$|\psi_{\text{seed}}\rangle = \sum_{n=1}^{N} c_n e^{i(\omega_n t + \phi_n)} |p_n, n\rangle \quad (22.13)$$

Here:

- The equation describes the quantum state of a seed configuration as a superposition of N quantum states, and by representing the fourfold aspect encoded as amplitude (power), frequency (knowledge), phase (harmony), polarization (presence), adheres to the QIQD framework.
- $e^{i(\omega_n t+\phi_n)}$: encodes the dynamic evolution of the n-the quantum state.
- c_n: Captures the strength of contribution (amplitude) for the n-th state.
- ω_n: Represents the vibrational or energetic aspect (frequency) of the seed's properties.
- ϕ_n: Encodes the relational and temporal dynamics of the state.
- p_n: Adds the spatial or directional orientation of the state, making the property multidimensional.
- $|n\rangle$ represents the nth eigenstate.
- $|p_n\rangle$ represents the polarization vector associated with the n-th basis state and is an independent property of each quantum state.

Dynamic MAR interaction would leverage (22.14), *Dynamic Interaction of Properties (MARs)*, that describes how the energetic imprints of different rings evolve over time, influenced by their interactions:

$$|\Psi_{\mathrm{MAR}}(t)\rangle = \prod_{j=1}^{M} \mathcal{T}\left[\mathrm{e}^{-i\int_{t_0}^{t} H_j(t',c_j,\omega_j,\phi_j,p_j)\mathrm{d}t'} \right] \left|\Psi_j(t_0)\right\rangle \tag{22.14}$$

Here

- $|\Psi_{\mathrm{MAR}}(t)\rangle$: Represents the combined state of all concentric rings within a Mixed Atom Ring (MAR) at time t.
- M: Number of rings in the MAR, each encoding a distinct seed or interaction.
- $c_j, \omega_j, \phi_j, p_j$: Amplitude, frequency, phase, and polarization of the jth ring's quantum property.
- $H_j\big(t', c_j, \omega_j, \phi_j, p_j\big)$: Time-dependent Hamiltonian of ring j, encoding the energetic interactions within that ring and between neighboring rings.
- $\mathcal{T}\left[\mathrm{e}^{-i\int_{t_0}^{t} H_j(t',c_j,\omega_j,\phi_j,p_j)\mathrm{d}t'} \right]$: The time-ordering operator ensures the evolution of states accounts for time-dependent changes in the Hamiltonian (system energy function).
- $\left|\Psi_j(t_0)\right\rangle$: Initial state of ring j at time t_0.

STEG-Fabric adjustment leverages the STEG-Fabric's adaptive topology control (22.15). This feature enables the lattice structure of the fabric to dynamically adjust its connectivity based on the ongoing computational demands. For instance, if certain qibits become pivotal in processing or encoding specific properties, the fabric's topology is modified to enhance these connections, thereby elevating the roles of these active qibits. In contrast, qibits that are less active might see a reduction in their connectivity, effectively modulating the network to optimize computational efficiency and responsiveness.

$$C_{ij}(t) = \alpha_{ij}(t).\, e\left(-\frac{\left|\vec{r}_i(t) - \vec{r}_j(t)\right|}{\xi}\right) \tag{22.15}$$

Here

- $C_{ij}(t)$ represents the strength of connectivity between qibits i and j at time t.
- $\alpha_{ij}(t)$ is a factor adjusting for the activity level of the qibits.
- $\vec{r}_i(t)$ and $\vec{r}_j(t)$ are their positions in the lattice.
- ξ is a characteristic length scale that modulates connectivity based on qibit proximity and activity.

Resonance detection (22.16) will identify and amplify coherence between energetic imprints generated by the device and target energetic imprints. Such resonance serves as an indicator of new structure.

$$R\left(\psi_{\text{system}}, X_x\right) = \sum_k \int \psi_{\text{system}}(t) X_x^*(t) A_k e^{-i(\omega_k t + \phi_k)} p_k \, dt \tag{22.16}$$

Here:

- A_k: Amplitude factor associated with the kth frequency component. A higher amplitude indicates a stronger alignment between the system's state, and the generated seed at this frequency.
- ϕ_k: Phase shift of the kth component.
- p_k: Polarization factor for the kth component, reflecting the vector orientation of the quantum state. Ensures that resonance takes into account vector alignment.
- ω_k: Frequency of interest for the kth detected resonance component.
- $\psi_{system}(t)$: The time-evolved quantum state of the system.
- $X_x^*(t)$: The complex conjugate of the generated quantum-functional seed. This ensures that the inner product between the states captures the alignment in both magnitude and phase.

Hamiltonian optimization leveraging (22.17), enables the QIQD Processing Unit (QIQD-PU) to dynamically modify the system's energy configuration to optimize interactions and maintain coherence.

$$H(t) = H_0 + \sum_{i=1}^{k} \gamma_i \widehat{P}_i(c_i, \omega_i, \phi_i, p_i) + \sum_{j=1}^{M} \delta_j \widehat{P}_j\left(c_j, \omega_j, \phi_j, p_j\right) + \kappa(t) \tag{22.17}$$

Here

- $H(t)$: The time-dependent Hamiltonian governing the quantum system's behavior.
- H_0: Baseline Hamiltonian representing the system's inherent energy landscape.
- $\widehat{P}_i, \widehat{P}_j$: Operators associated with seed and MAR properties, respectively.
- γ_i, δ_j: Coefficients that adjust the influence of seed and MAR properties on the Hamiltonian.

- $\kappa(t)$: A feedback-driven term that introduces corrections based on real-time monitoring of coherence and system stability.

22.4 Further Reading

- "Complexity: The Emerging Science at the Edge of Order and Chaos" by M. Mitchell Waldrop: This book provides an introduction to the field of complex systems, exploring how emergent behavior arises in various domains, from ant colonies to economies.
- "Thinking in Systems: A Primer" by Donella Meadows: This book offers a concise introduction to systems thinking, emphasizing the interconnectedness of components and the dynamics of feedback loops. This is crucial for understanding how CAS function and evolve.
- "A Different Universe: Reinventing Physics from the Bottom Down" by Laughlin, Robert B.: Discusses "emergence" as a phenomenon where collective behaviors overshadow lower-level rules.
- "The Emergence of Everything: How the World Became Complex" by Morowitz, Harold J.: Traces the hierarchical emergence of complexity from physics to life.
- "Mind in a Physical World: An Essay on the Mind–Body Problem and Mental Causation" by Kim, Jaegwon: From a philosophy of mind perspective, explores "supervenience" and emergent properties.

References

1. Lewin R (1999) Complexity: life at the edge of chaos. University of Chicago Press, Chicago
2. Prigogine I (1997) The end of certainty: time, chaos, and the new laws of nature. Free Press, New York
3. Malik P (2024) Sensing the emergence of new structure in matter and life and its impact on physics, chemistry, and biology, in a light-based quantum computational model. In: Yang XS, Sherratt S, Dey N, Joshi A (eds) Proceedings of ninth international congress on information and communication technology. ICICT 2024 2024. Lecture notes in networks and systems, vol 1011. Springer, Singapore. https://doi.org/10.1007/978-981-97-4581-4_7
4. Sayrin C et al (2011) Real-time quantum feedback prepares and stabilizes photon number states. Nature 477:73–77. https://doi.org/10.1038/nature10376
5. Schrödinger E (1992) What is life? With mind and matter and autobiographical sketches. Cambridge University Press, Cambridge, 1992 (reprint of 1944 original).
6. Nielsen MA, Chuang IL (2010) Quantum computation and quantum information. Cambridge University Press, Cambridge
7. Carleo G, Troyer M (2017) Solving the quantum many-body problem with artificial neural networks. Science 355(6325):602–606. https://doi.org/10.1126/science.aag2302
8. Georgescu IM, Ashhab S, Nori F (2014) Quantum simulation. Rev Mod Phys 86(1):153–85. https://doi.org/10.1103/RevModPhys.86.153

Chapter 23
A QIQD Nano-based Device

Abstract This chapter introduces a QIQD-based nano-device with the ability to detect, sense, and generate quantum-functional properties, effectively bridging the gap between the quantum realm and tangible reality. This device operates by first detecting the Organizing Quantum Energy (OQE) within a system, identifying the specific quantum-functional seed that drives its organizational structure. It achieves this through a sophisticated interplay of quantum-field sensors, resonance gates, and seed generators, capturing intricate quantum patterns and isolating the dominant seed through a feedback-controlled system. Beyond detection, the device also senses emergent structures, tracking the evolution of new quantum properties in real time. This dynamic capability is enabled by the STEG-Fabric, a programmable lattice structure with adaptive topology control, along with temporal dynamics detectors and specialized components like Mixed Atom Rings (MARs) and Pure Property Quantum Dots (PPQDs). Furthermore, the device can generate entirely new energetic imprints, encoding, modulating, and synthesizing them using the STEG-Fabric as a medium. Qibits are initialized with specific spectral characteristics, and dynamic modulation circuits ensure their coherence and adaptability. Programmable matter synthesis is facilitated by gates like PP-Adders and PP/PAR-Subtractors, while the Adaptive Hamiltonian Controller maintains stability. These diverse functionalities are seamlessly integrated through a central QIQD Processing Unit (QIQD-PU), which orchestrates operations and manages coherence across multiple layers. Housed within a nanoscale platform designed to minimize decoherence, this device holds potential for applications in advanced materials, quantum computing, and adaptive systems.

A QIQD nano-based device is envisioned [1],[1] having three key operating principles: (i) detecting OQE in any structure; (ii) sensing new and emerging energetic imprints; and (iii) generating new energetic imprints that can be programmed into matter via a removable STEG-Fabric.

---

[1] The original idea of a 'nano-cyborg' appeared in an IEEE article [1] but has been further developed based on accumulated QIQD insights as developed in this book.

P. Malik, *Pioneering New Avenues in Quantum Technology*, Studies in Smart Technologies, https://doi.org/10.1007/978-981-96-5463-5_23

There are many applications that such a device will provide the foundation for, and this chapter will summarize these three principles via mathematical equations, and then extrapolate some representative use cases.

23.1 Summary of QIQD Hardware and Operating Principle Mathematics

Leveraging the mathematics described in the book, three summary QIQD hardware and equations are suggested that embody the three key operating principles envisioned for a QIQD nano-device.

To detect organizing quantum energy (OQE), the nano-device relies on a system of quantum-field sensors, resonance gates, and quantum-functional seed generators as discussed in Chap. 22. These sensors are highly sensitive to coherence, phase relationships, and amplitude variations within quantum states. Their purpose is to capture the intricate quantum patterns that define the organizational structure of a system, providing the raw data needed for analysis. Resonance gates align the detected quantum states with archetypal seeds encoded in the system. These gates employ advanced quantum optics and nanophotonic circuits to amplify weak signals and confirm resonance through phase and polarization alignment. The seed generators, drawing on tunable quantum fields, produce archetypal quantum-functional patterns representing fundamental properties like Knowledge, Presence, Power, and Harmony. These components are integrated into a feedback-controlled system that iteratively refines detection accuracy and ensures the isolation of dominant seeds driving the system's organization.

Equation 23.1, *Isolation of Dominant Seed*, embodies the principle to do with detecting OQE in any structure by isolation of the dominant seed:

$$\mathcal{O}_{\mathrm{OQE}}(X) = \arg\max_{k}\left[\int \Gamma(X, \Phi_k) A_k \mathrm{e}^{i(\omega_k t+\phi_k)} \vec{p}_k \vec{p}_x \mathrm{d}t\right] \tag{23.1}$$

Here:

- Input State (X): Represents the system's current quantum-functional properties, and encodes dynamic interactions between quantum properties like frequency, amplitude, phase, and polarization.
- Seeds (Φ_k): Archetypal quantum basis functions representing foundational organizational properties (e.g., Knowledge, Presence, Power, Harmony). Predefined in QIQD hardware (e.g., generators and qibits), these seeds act as templates for recognizing quantum-functional order.
- Overlap Function ($\Gamma(X, \Phi_k)$): Measures the coherence between the input state X and a specific seed Φ_k. Indicates how well the seed aligns with the organizational structure of the target system.

- Resonance Weight ($A_k e^{i(\omega_k t+\phi_k)}$): Modulates the seed's contribution to the system's organization based on its amplitude (A_k), frequency (ω_k), and phase (ϕ_k). Describes how strongly a seed is expressed in the system.
- Integration Over Time ($\int \ldots dt$): Captures the temporal evolution of resonance between X and Φ_k, ensuring the analysis includes dynamic interactions and emergent properties.
- Maximization ($\arg\max_k$): Selects the seed Φ_k with the highest resonance, identifying it as the dominant driver of the system's organization.
- Polarization:

  (a) $\vec{p}_k$: Polarization vector of the k-th seed, representing its orientation in a multi-dimensional space.
  (b) $\vec{p}_x$: Polarization vector of the input quantum-functional state X.
  (c) $\vec{p}_k\vec{p}_x$: The dot product measures the alignment (cosine of the angle) between the polarization vectors of the seed and the input state.

The module for sensing emergent structures is a dynamic and adaptive subsystem designed to track the evolution of new quantum properties in real time. It builds on the foundation of STEG-Fabric, a programmable lattice structure capable of embedding qibits with encoded energetic imprints. This fabric is equipped with adaptive topology control, allowing it to modify its connectivity and respond dynamically to emergent quantum states. The sensing hardware also incorporates temporal dynamics detectors, which monitor changes in the system's coherence and structural complexity over time. Mixed Atom Rings (MARs) and Pure Property Quantum Dots (PPQDs) further enhance this module by enabling the capture and manipulation of emergent quantum configurations.

Equation 23.2, *Sensing New Energetic Imprints* embodies the principle to do with sensing emergent imprints from detected quantum states in its most generalized form:

$$Q_{\text{STEG}} = \left(S' \wedge T' \wedge E' \wedge G'\right)\Theta\left(Q_{\text{dynamic}}\right) \tag{23.2}$$

Here:

- STEG Quantization Dynamics (S', T', E', G'): These terms represent manifestations of Space (S'), Time (T'), Energy (E'), and Gravity (G')—the four-dimensional code that accompanies any quantum object—and in the province of quantum systems, are also recognizable via the spectral properties of frequency (knowledge), power (amplitude), phase (harmony), and polarization (presence) characteristic of any material quantum object.
- Logical Conjunction ($\wedge$): This operator signifies the interdependence of S', T', E', G'. A new energetic imprint only emerges fully when all four properties are coherently aligned, reflecting the unified action of space, time, energy, and gravity.
- Dynamic Evolution ($(\Theta(Q_{\text{dynamic}}))$: Q_{dynamic} represents the dynamic state of the quantum system, encompassing its continuous interactions and transformations. Θ is a function encoding the temporal evolution and feedback from the quantum system into the STEG quantizations. Θ can be thought of as a mechanism that

keeps STEG quantizations aligned with the evolving quantum state making them a reliable basis for detecting and encoding new energetic imprints.

To generate programmable matter, the nano-device integrates a specialized system for encoding, modulating, and synthesizing energetic imprints. At the heart of this system is the STEG-Fabric, which serves as a medium for encoding quantum properties and fostering their dynamic evolution. The encoding process involves initializing qibits with spectral characteristics—frequency, amplitude, phase, and polarization—that map to desired energetic imprints. Dynamic modulation circuits enable real-time adjustments to these properties, ensuring that the encoded states remain coherent and adaptable to external stimuli. The programmable matter synthesis is facilitated by gates such as PP-Adders and PP/PAR-Subtractors, which combine or refine quantum properties to produce new configurations. Additionally, the Adaptive Hamiltonian Controller embedded in the device ensures the stability and coherence of generated energetic configurations by dynamically tuning the system's energy landscape.

Three Eqs. (23.3–23.5) summarize the gist of programming matter. Equation (23.3), *Qibit Encoding*, is summarized as:

$$|\psi\rangle = \sum c_n |f_n, a_n, \phi_n, p_n\rangle \tag{23.3}$$

Here:

- f_n, a_n, ϕ_n, p_n: Frequency, amplitude, phase, and polarization of the n-th spectral component.
- c_n: Coefficient weighting each property's contribution.

Equation (23.4), *Dynamic Modulation*, is depicted as:

$$\Delta|\psi\rangle = \sum \delta c_n |f_n + \Delta f, a_n + \Delta a, \phi_n + \Delta\phi, p_n + \Delta p\rangle \tag{23.4}$$

Here encoded properties are adjusted dynamically based on feedback.

Equation (23.5), *Emergent Property Synthesis*, is depicted as:

$$P_{\text{emergent}} = f\left(\sum \alpha_i P_{\text{seed},i} + \beta_j P_{\text{MAR},j}\right) \tag{23.5}$$

Here:

- $P_{\text{seed},i}$: Represents the contributions of foundational seeds encoded within the qibits.
- $P_{\text{MAR},j}$: Represents properties derived from Mixed Atom Rings (MARs), which encode complex interactions and emergent dynamics.
- α_i, β_j: Coefficients weighting the contributions of seeds and MARs to the emergent property.
- $f()$: A transformation function synthesizing these properties into a new configuration.

To synthesize these functionalities into a unified nano-device, the hardware components are interconnected through a central QIQD Processing Unit (QIQD-PU). This unit orchestrates operations across sensors, gates, and fabric layers. It leverages advanced quantum computing techniques, including machine learning-assisted pattern recognition, to dynamically adjust and optimize the device's performance. The QIQD-PU also manages multi-layered coherence, ensuring that the interactions remain harmonized. The entire system is housed within a nano-scale platform built from materials designed to minimize decoherence and thermal noise, such as superconducting circuits and photonic resonators.

If the envisioned QIQD nano-based device needs to unify its three functions so that detection informs sensing, sensing refines synthesis, and synthesis validates detection, Eq. (23.6), *Unified Nano-device Equation*, would be as follows:

$$Q_{\text{nano}}(t) = \arg\max_{k}\left[\int \Gamma(X, \Phi_k)\mathrm{A}_k \mathrm{e}^{i(\omega_k t+\phi_k)}\vec{\mathrm{p}}_k\vec{\mathrm{p}}_x \mathrm{d}t\right]$$
$$\Theta\left(Q_{\text{dynamic}}\right)\left(\sum \alpha_i P_{\text{seed},i} + \beta_j P_{\text{MAR},j}\right) \tag{23.6}$$

By combining these specialized components into a cohesive nano-device, the QIQD framework achieves a powerful, adaptive system capable of detecting, sensing, and generating quantum-functional properties. This device embodies the principles of dynamic interaction and emergent organization, serving as a foundational tool for applications in advanced materials, quantum computing, and adaptive systems.

23.2 Envisioning Representative Use Cases

The envisioned QIQD nano-based device, capable of detecting Organizing Quantum Energy (OQE), sensing new energetic imprints, and programming new configurations within its STEG-Fabric, holds transformative potential across a range of fields. Its envisioned ability to operate at the quantum-functional level of properties and interactions suggests uses that span health, energy, materials science, and beyond, each embodying a narrative of innovation.

In healthcare, this technology could revolutionize diagnostics and treatments by detecting the most subtle energetic patterns that precede physical manifestations of diseases. Imagine a device that identifies the energetic imbalances within cells signaling the earliest stages of cancer or neurodegeneration, long before conventional tests can detect them. By capturing these imprints, the device would not only diagnose but also introduce precise energetic corrections, fostering cellular healing. Personalized medicine would take on a new meaning as this device could tailor interventions at the quantum level to resonate with the unique energetic signature of each individual, enabling therapies that are not just effective but profoundly integrative.

The material sciences stand to be equally transformed. This device could create programmable matter, engineering combinations of properties that nature alone cannot provide. For example, room-temperature superconductors, which remain a scientific dream, could become a tangible reality. These materials, capable of conducting electricity with perfect efficiency at ambient temperatures, would eliminate energy loss in transmission, promoting ultra-efficient power grids and revolutionary transportation systems. Similarly, the QIQD device could stabilize metallic hydrogen, unlocking new frontiers in energy storage and propulsion, or design artificial photosynthetic materials that far surpass the efficiency of natural systems, directly converting sunlight into clean energy.

In energy and sustainability, the QIQD device could introduce unmatched efficiency and ecological balance. By creating materials that dissolve waste harmlessly or that adapt dynamically to environmental conditions, the device could address long-standing challenges of pollution and resource management. The ability to tailor energetic imprints would also extend to designing more resilient solar cells, optimizing the capture and conversion of sunlight to electricity in ways previously unimaginable.

Quantum computing and artificial intelligence (AI) also find new dimensions with the QIQD framework. The device's capacity to detect and refine energetic imprints could stabilize qibits, significantly enhancing the reliability and power of quantum computers. Moreover, by embedding properties such as coherence and adaptability into its computational framework, the device could give rise to emergent AI systems that behave more like organic intelligences, dynamically learning and evolving in complex environments.

Aerospace and defense applications would benefit from the creation of materials that withstand extreme conditions. The QIQD device could engineer alloys that remain stable under the intense radiation and thermal stress of outer space or design sensors capable of detecting the most minute changes in environmental parameters, enhancing navigation and safety.

Lastly, there are significant implications for broader systems, from urban planning to global networks. The device could lead to the development of smart materials that adapt in real time to changing environments, such as self-cooling structures in extreme heat or self-healing infrastructure in earthquake zones. Networks of these materials, all programmed and controlled at the quantum level, would redefine our interaction with the built and natural worlds, creating cities and systems that are not only intelligent but harmoniously integrated with their surroundings.

Through these uses, the QIQD device promises to merge the abstract elegance of quantum dynamics with the tangible needs of human life, reshaping the possibilities of science and technology while fostering a future of adaptability, efficiency, and profound innovation. Each application would reflect the unique interplay of presence, power, knowledge, and harmony that the device embodies, making it a cornerstone in QIQD-based advancement.

23.3 Further Reading

- "Quantum Engineering: Theory and Design of Quantum Coherent Structures" by A. M. Zagoskin: This book provides a comprehensive overview of quantum engineering principles, covering topics like superconducting circuits, quantum dots, and nanophotonics, which are relevant to the construction of the proposed nano-device.
- "Principles of Nano-Optics" by Lukas Novotny and Bert Hecht: This text delves into the fundamentals of light-matter interactions at the nanoscale, exploring concepts like near-field optics and plasmonics, which could be relevant to the device's sensing and manipulation capabilities.
- "Nanoelectronics and Nanosystems: From Transistors to Molecular and Quantum Devices" by Karl Goser, Peter Glösekötter, and Jan Dienstuhl: This book explores the world of nanoelectronics, covering various nanoscale devices and their applications, providing context for the potential functionalities of the QIQD nano-device.
- "Quantum Materials: Experiments and Theory" edited by Subir Sachdev and Bernhard Keimer: This book explores the world of quantum materials, where quantum phenomena give rise to exotic properties like superconductivity and topological phases. This is relevant to the potential applications of the QIQD device in creating novel materials.
- "Quantum Computing in the NISQ Era and Beyond" by Preskill, John: Discusses near-term quantum hardware challenges and the path toward more sophisticated architectures.
- "Quantum Error Correction for Beginners." By Devitt, Simon J., Ashley D. Greentree, and Lloyd C. L. Hollenberg: Reveals how error correction can maintain coherence.

Reference

1. Malik P (2022) Envisioning a light-based quantum-computational nano-cyborg. In: 2022 IEEE International IOT, Electronics and Mechatronics Conference (IEMTRONICS), Toronto, ON, Canada, pp 1–8. https://doi.org/10.1109/IEMTRONICS55184.2022.9795762

Chapter 24
Unifying Quantum Seeds and Spacetime

Abstract This chapter proposes a conceptual bridge between general relativity and quantum dynamics by embedding aspects of the quaternary framework (QIQD) directly into the Einstein Field Equations (EFE). Building on the classical curvature–energy relationship outlined by EFE, the chapter introduces an additional divergence term, ∇·(Seed Properties), representing four quantum-functional aspects (Presence, Power, Knowledge, Harmony). By coupling these "seed dynamics" to traditional stress-energy terms, a modified EFE that potentially unifies macroscopic spacetime curvature with microscopic quantum effects is obtained. The four seed properties collectively influence curvature, temporal evolution, vacuum-energy balance, and local material density, offering a novel lens into phenomena such as quantum-assisted gravitation and emergent spacetime structures. This framework also points to far-reaching technological applications: from more stable quantum computing and advanced superconductors to quantum-tuned navigation and AI systems. The resulting QIQD–EFE synthesis opens up new directions in both theoretical physics and engineering, suggesting that the interplay of quantum seed properties and spacetime geometry can be harnessed for breakthroughs spanning cosmology, quantum gravity, and beyond.

This chapter begins with a hi-level synthesis to suggest how spacetime can potentially be influenced at the micro, quantum levels [1, 2], particularly through the deterministic dynamics of seeds as framed in the Quaternary Interpretation of Quantum Dynamics (QIQD). This insight arises from the connection between the Einstein Field Equation (EFE) [3] and the flux or variation (divergence) of seed properties (Presence, Power, Knowledge, Harmony).

In Eq. (24.1), EFE, we see how space–time–energy–gravity are intimately connected:

$$R_{\mu\upsilon} - \frac{1}{2}Rg_{\mu\upsilon} + \wedge g_{\mu\upsilon} = \frac{8\pi G}{c^4}T_{\mu\upsilon} \tag{24.1}$$

Specifically, the left side of the equation describes the geometry of spacetime, while the right side represents the distribution of matter and energy.

P. Malik, *Pioneering New Avenues in Quantum Technology*, Studies in Smart Technologies, https://doi.org/10.1007/978-981-96-5463-5_24

Note that the '$\mu\upsilon$' are the following [4]:

- $g_{\mu\upsilon}$ (Metric tensor): encodes the geometric and causal structure of spacetime.
- $T_{\mu\upsilon}$ (Stress-energy tensor): Describes the distribution of energy, momentum, and stress within spacetime.
- $R_{\mu\upsilon}$ (Ricci tensor): A contraction of the Riemann curvature tensor, representing part of the spacetime curvature.

This equation captures the interplay between spacetime geometry and classical energy-matter distributions, but it leaves out the subtler quantum dynamics that shape and influence spacetime on finer scales.

24.1 Integrating QIQD with EFE

Connecting the terms in EFE to QIQD begins to construct a view of how subtler quantum dynamics influence spacetime:

- $R_{\mu\upsilon}$: Ricci Tensor captures how the shape of spacetime is influenced by mass and energy distributions. It encodes the spatial curvature of spacetime, and in QIQD maps to Knowledge (space).
- $-{}^{1}/_{2}Rg_{\mu\upsilon}$: The term governs how the overall curvature impacts spacetime's temporal structure. It represents a dynamic scalar quantity describing the evolution of spacetime as a whole and is associated with Power (time) in QIQD.
- $\wedge g_{\mu\upsilon}$: Cosmological constant term represents the uniform curvature contributed by vacuum energy, and is linked to Harmony, balancing spacetime's geometry against perturbations.
- $\frac{8\pi G}{c^4}T_{\mu\upsilon}$: Presence (material density) influences $T_{\mu\upsilon}$, representing the energy-matter distribution.

Recalling (4.8) the equation for a seed:

$$\text{Seed } = Xa + \sum_{i=0}^{n} Y_i b_i \quad \text{where} \begin{bmatrix} X, Y \in [S_{\text{Pr}}, S_{\text{Po}}, S_{\text{K}}, S_{\text{H}}] \\ a, b \text{ are integers; } a > b \end{bmatrix} \tag{4.8}$$

Combining (24.1) and (4.8) yields Eq. (24.2), *EFE Seed Dynamics*:

$$R_{\mu\upsilon} - {}^{1}/_{2}Rg_{\mu\upsilon} + \wedge g_{\mu\upsilon} \sim \frac{8\pi G}{c^4}\underset{\mu\upsilon}{\breve{T}}$$
$$+ \nabla \cdot \text{Seed Dynamics(Presence, Power, Knowledge, Harmony)} \tag{24.2}$$

In (24.2) T becomes T' which implies that:

$$\frac{8\pi G}{c^4}T_{\mu\upsilon} = \frac{8\pi G}{c^4}\underset{\mu\upsilon}{\breve{T}}$$

$$+\nabla \cdot \text{Seed Dynamics}(\text{Presence, Power, Knowledge, Harmony})$$

The implication is that the divergence operator (∇) which represents the flux or variation of foundational quantum properties—Presence, Power, Knowledge, and Harmony—across spacetime, contributes to the energy-mass dynamic represented by the right side of the equation. These properties arise from QIQD's mathematical framework, which views reality as a multi-layered system where quantum seeds encode the essence of physical and informational dynamics. Presence relates to material density and localizes the influence of energy and mass, Power governs temporal dynamics, Knowledge modulates spatial coherence and curvature, and Harmony ensures stability and balance in spacetime.

Each term in the EFE can be associated with the flux or divergence of a specific aspect of seed dynamics:

- $R_{\mu\upsilon} \sim \nabla$. (Knowledge Seed Contribution): The Ricci tensor captures the divergence of Knowledge-related seeds reflecting influence on spatial coherence and curvature.
- $-{}^{1}/{}_{2}Rg_{\mu\upsilon} \sim \nabla \cdot$(Power Seed Contribution): The trace term captures the divergence of Power-related seeds, reflecting their influence on temporal dynamics and global spacetime evolution.
- $\wedge g_{\mu\upsilon} \sim \nabla \cdot$(Harmony Seed Contribution): Harmony seeds contribute to global balance, ensuring spacetime coherence and stability, represented by the cosmological constant term.
- $T_{\mu\upsilon} \sim \nabla \cdot$(Presence Seed Contribution): Presence seeds ground spacetime geometry through localized material-energy density.

Whereas (24.2) represents the point of view of quantum objects operating at the B-layer in UQO where mass-energy (as Presence) is the anchor point, so to speak, it is possible to also take the point of view of precipitation from the meta-layers (M3, M2, M1) to B as has been done in Eqs. (10.1–10.4). These represent the quantum object code for a simple hypothetical atom showing how the equation for space evolution (S_e), is central for understanding the dynamics of time evolution (T_e), energy evolution (E_e), and gravity evolution (G_e) that exist in unison as the four-dimensional code for any quantum object. Reproducing:

$$S_e = h\left(\vec{X} + \sum_{i=0}^{n} \vec{Y}_i\right) \cdot \Theta(\text{seed}) \tag{10.1}$$

$$T_e = h\left(T_o + \tilde{\vec{X}} + \sum_{i=0}^{n} \tilde{\vec{Y}}_i\right) \cdot \Theta(S_e) \tag{10.2}$$

$$E_e = h\left(E_o + \tilde{\vec{X}} + \sum_{i=0}^{n} \tilde{\vec{Y}}_i\right) \cdot \Theta(S_e) \cdot (1 + \eta \cdot \mathcal{M}(S_k)) \tag{10.3}$$

$$G_e = h\left(G_o + \tilde{\vec{X}} + \sum_{i=0}^{n} \tilde{\vec{Y}}_i\right) \cdot \Theta(S_e) \cdot \left(1 + \eta \cdot \mathcal{U}\left(\vec{X}, \left\{\vec{Y}_i\right\}\right)\right) \tag{10.4}$$

The implication here is that seeds originating at M1 to set up 'space' [(10.1) related to $R_{\mu\upsilon}$] then mature at B through specific time dynamics [(10.2) related to $-{}^{1}/{}_{2}Rg_{\mu\upsilon}$], materialization dynamics [(10.3) related to $T_{\mu\upsilon}$], and gravity dynamics [(10.4) related to $\wedge g_{\mu\upsilon}$].

24.2 Further Elaboration of Seed Dynamics

By introducing '∇·Seed Dynamics,' Eq. (24.2) acknowledges the dual contribution of macroscopic classical sources and microscopic quantum effects. This unification reflects the emergent nature of macroscopic phenomena from underlying quantum dynamics. The divergence term [5] allows quantum seed properties to manifest as localized effects on spacetime curvature while maintaining their global influence, ensuring consistency across scales [6]. Note that the validity of combining macroscopic and microscopic contributions in a single equation lies in the fundamental principle of scale bridging. In physics, unified descriptions often integrate phenomena from different scales, as seen in Maxwell's equations for electromagnetism [7] or the Navier–Stokes equations for fluid dynamics [8]. Similarly, the '∇·Seed Dynamics' term mirrors this tradition, incorporating quantum-level information without disrupting the macroscopic integrity of the EFE.

The introduction of '∇·Seed Dynamics' transforms EFE into a tool for exploring both classical and quantum contributions to spacetime geometry. The equation potentially describes phenomena like quantum fluctuations in spacetime, emergent gravitational effects [2], or even the interplay of spacetime with quantum systems like QIQD-engineered materials. It serves as a conceptual bridge between general relativity and quantum dynamics, possibly offering a step toward a unified theory of quantum gravity.

Mathematically, the divergence operator (∇·) captures the flux of quantum properties across spacetime, connecting local variations to global effects. For example, the term may be expressed as Eq. (24.3), *Interplay of Properties over Spacetime*:

$$\nabla \cdot (\text{Seed Properties}) = \frac{\partial \rho_{\text{Pr}}}{\partial x} + \frac{\partial \rho_{\text{Po}}}{\partial y} + \frac{\partial \rho_{\text{K}}}{\partial y} + \frac{\partial \rho_{\text{H}}}{\partial z} \tag{24.3}$$

where ρ_{Pr}, ρ_{Po}, ρ_{K}, ρ_{H} denote the densities of Presence, Power, Knowledge, and Harmony, respectively. This formulation highlights the dynamic interplay of these properties across spacetime coordinates, reflecting both local and non-local influences:

- ρ_{Pr}: Primarily linked to spatial distributions (e.g., mass or material density). Its variation with x reflects how presence changes along a spatial dimension.

- ρ_{Po}: Relates to temporal dynamics. Its variation with t captures how power evolves over time.
- ρ_{K}: Encodes coherence and structure in space. Its dependence on y reveals how spatial configurations change across one dimension of space.
- ρ_{H}: Balances and stabilizes interactions across spacetime. Its variation with z reflects spatial adjustments that support systemic stability.

In this framework, x, t, y, z correspond to the spacetime coordinates over which these properties are distributed, making the association a natural consequence of dimensional relevance.

The divergence operator ∇ measures the rate at which a field (in this case, seed properties) expands or contracts within a region. In mathematical terms, it computes the partial derivatives of each field component along the respective spatial or temporal coordinate. The association between the denominator and the numerator arises from how these derivatives are computed:

- $\frac{\partial \rho_{\mathrm{Pr}}}{\partial x}$: Rate of change of the property ρ_{Pr} (Presence) along the x-axis.
- $\frac{\partial \rho_{\mathrm{Po}}}{\partial y}$: Rate of change of ρ_{Po} (Power) with respect to time, reflecting temporal variation.
- $\frac{\partial \rho_{\mathrm{K}}}{\partial y}$: Rate of change of ρ_{K} (Knowledge) along the y-axis, capturing structural changes.
- $\frac{\partial \rho_{\mathrm{H}}}{\partial z}$: Rate of change of ρ_{H}: : (Harmony) along the z-axis, encoding spatial adjustments.

This differentiation aligns each quantum property with the dimension where its effects are most directly observable or measurable.

24.3 Technological Advancements

The integration of the Quaternary Interpretation of Quantum Dynamics (QIQD) with the modified Einstein Field Equations (EFE) introduces a revolutionary framework that unifies quantum seed properties—Presence, Power, Knowledge, and Harmony—with spacetime geometry. This unification bridges the quantum and macroscopic realms, transforming our understanding of reality's structural dynamics and offering profound implications for technological advancements.

Quantum computing, for example, is poised to benefit from this framework by embedding quantum seed dynamics directly into spacetime fabric [5]. Harmony, through its stabilizing influence on spacetime curvature, could enhance qibit coherence, while Power seeds dynamically govern temporal evolution, optimizing algorithmic processes.

In energy systems, the relationship between quantum seeds and spacetime flows introduces the potential for quantum-optimized grids and superconductors. By leveraging the coherence encoded in Knowledge and the localized energy densities of Presence, these systems can minimize losses and revolutionize energy storage, potentially achieving breakthroughs like room-temperature superconductivity [9].

The framework also provides a unique lens for cosmological simulations. The divergence of quantum seeds, as described in Eqs. (24.2) and (24.3), may be leveraged to model phenomena such as cosmic inflation [10] where Harmony stabilizes vacuum fluctuations, or quantum-corrected black holes [11] where Power seeds influence singularity dynamics. This may allow for a more complete understanding of the universe's evolution and structure.

Precision sensing technologies are another frontier, where systems grounded in the dynamics of QIQD-EFE could detect spacetime distortions with unparalleled accuracy [12]. By measuring the flux of Presence and Knowledge seeds, quantum gravitational sensors and OQE-detection systems could redefine geophysics, navigation, and space exploration. Similarly, the manipulation of emergent spacetime structures using Power and Harmony seeds may lead to applications like stable wormholes or quantum-based energy storage.

In biology, the quantum underpinnings of cellular processes could be illuminated through the dynamics of seed interactions with spacetime. Knowledge seeds, governing spatial coherence, might guide programmable DNA design [13], while Harmony could stabilize synthetic biological systems for enhanced functionality.

Artificial intelligence stands to be profoundly impacted by this unification. Quantum-informed AI algorithms [14] could draw directly from the deterministic dynamics of seed properties to model complex systems across scales. Unified physics simulators, leveraging the framework's multi-layered interactions, could predict behaviors in systems that straddle quantum and classical domains, from black holes to high-energy particles.

The integration of QIQD and EFE also opens the door to quantum-tuned communication and navigation systems. By embedding Harmony and Power seeds into spacetime-resilient networks, these systems could achieve unmatched signal stability and precision, enabling ultra-accurate quantum GPS systems.

Finally, the framework offers a transformative step forward in the pursuit of quantum gravity. By explicitly incorporating quantum seeds into spacetime curvature, it provides a testable bridge between quantum mechanics and general relativity. This opens new experimental possibilities, from probing the emergence of spacetime structures to exploring the interplay of vacuum energy and curvature dynamics. The QIQD-EFE framework not only deepens our understanding of the universe but also lays the groundwork for innovations that can harness the fundamental principles of existence. By uniting quantum and macroscopic dynamics, it transforms the way we understand, engage with, and ultimately shape the cosmos.

24.4 Further Reading

- "100 Years of General Relativity: Space–Time Structure—Einstein and Beyond" by Ashtekar, Abhay, ed.: A set of essays reflecting on the century of progress since Einstein, addressing the open challenge of merging quantum mechanics and spacetime curvature—conceptually akin to QIQD-EFE.
- "The Large-N Limit of Superconformal Field Theories and Supergravity" by Maldacena, Juan: Key paper that introduced the AdS/CFT correspondence, hinting at deep connections between quantum field theories and gravitational descriptions—potentially relevant for bridging "seed properties" and spacetime in higher-dimensional frameworks.
- "The Road to Reality: A Complete Guide to the Laws of the Universe" by Penrose, Roger: An expansive overview of physics, from classical geometry to quantum field theory; helpful for situating new interpretations within a broader mathematical landscape.
- "The Black Hole War: My Battle with Stephen Hawking to Make the World Safe for Quantum Mechanics" by Susskind, Leonard: A more accessible introduction to black hole information paradoxes, holography, and how quantum effects must be reconciled with spacetime curvature—ties in with the notion of "quantum seeds" influencing singularities.
- "Particle Creation by Black Holes" by Hawking, Stephen W.: Shows how quantum fluctuations in curved spacetime can lead to observable phenomena (Hawking radiation)—a prototype for how "seed dynamics" might produce emergent gravitational effects.
- "The Problem of Time: Quantum Mechanics versus General Relativity" by Anderson, Edward: Examines the clash between timeless quantum wavefunctions and the dynamic flow of time in general relativity.
- "Decoding Reality: The Universe as Quantum Information" by Vlatko Vedral: Proposes that the whole of physical reality can be understood in terms of information and its processing.
- "Roadmap on Atomtronics: State of the Art and Perspective." By Amico, Luigi, et al.: Provides a detailed look at "atomtronics," or atomic circuits, which parallel the idea of Mixed Atom Rings (MARs) used in QIQD to encode emergent quantum states.
- "The Character of Physical Law" by Feynman, Richard P.: A classic series of lectures offering insights into how physical laws might reflect deeper symmetries or underlying structures—perhaps like the seed properties driving QIQD's "fourfold" approach.
- "Quantum Sensing and Precision Metrology" by Bressler, Shlomo, and Grégoire Ribordy, eds.: Offers developments in precision metrology—extending "quantum gravitational sensors" and "OQE-detection systems" that might measure subtle spacetime effects in QIQD-EFE.

- "The Embodied Mind: Cognitive Science and Human Experience" by Varela, Francisco J., Evan Thompson, and Eleanor Rosch: Though from a cognitive science perspective, it addresses how systemic coherence and feedback loops create emergent phenomena.
- "More Is Different." by Anderson, Philip W.: Classic argument on how higher-level phenomena emerge from simpler building blocks—a conceptual parallel to how "quantum seed dynamics" might feed into EFE at large scales.

References

1. Rovelli C (2004) Quantum gravity. Cambridge University Press, Cambridge
2. Kiefer C (2012) Quantum gravity, 3rd edn. Oxford University Press, Oxford
3. Einstein A (1916) The foundation of the general theory of relativity. Ann Phys 49:769–822
4. Misner CW, Thorne KS, Wheeler JA (1973) Gravitation. W. H. Freeman, San Francisco
5. Birrell ND, Davies PCW (1982) Quantum fields in curved space. Cambridge University Press, Cambridge
6. Laughlin RB, Pines D (2000) The theory of everything. Proc Natl Acad Sci 97(1):28–31. https://doi.org/10.1073/pnas.97.1.28
7. Jackson JD (1998) Classical electrodynamics, 3rd edn. Wiley, New York, NY
8. Landau LD, Lifshitz EM (1987) Fluid mechanics, 2nd edn. Butterworth-Heinemann, Oxford
9. Bardeen J, Cooper L, Schrieffer JR (1957) Theory of superconductivity. Phys Rev 108(5):1175–1204
10. Guth AH (1981) Inflationary universe: a possible solution to the horizon and flatness problems. Phys Rev D 23(2):347–356. https://doi.org/10.1103/PhysRevD.23.347
11. Hawking SW, Penrose R (1996) The nature of space and time. Princeton University Press, Princeton
12. LIGO Scientific Collaboration and Virgo Collaboration (2016) Observation of gravitational waves from a binary black hole merger. Phys Rev Lett 116(6):061102. https://doi.org/10.1103/PhysRevLett.116.061102
13. Davies P, Gregersen NH (Eds) (2010) Information and the nature of reality: from physics to metaphysics. Cambridge University Press, Cambridge
14. Briegel HJ (2012) On creative machines and the physical origins of freedom. Sci Rep 2:522. https://doi.org/10.1038/srep00522

The manufacturer's authorised representative in the EU is Springer Nature Customer Service Centre GmbH, Europaplatz 3, 69115 Heidelberg, Germany. If you have any concerns regarding our products, please contact ProductSafety@springernature.com

Printed and bound by CPI Group (UK) Ltd, Croydon, CR0 4YY
01/10/2026
02225934-0006